THE AVID
HANDBOOK

THE AVID HANDBOOK

Advanced Techniques, Strategies, and Survival Information for Avid Editing Systems

5th Edition

GREG STATEN
STEVE BAYES

Routledge
Taylor & Francis Group

LONDON AND NEW YORK

First published 2009 by Focal Press

Published 2022 by Routledge

4 Park Square, Milton Park, Abingdon, Oxon OX14 4RN
605 Third Avenue, New York, NY 10017

Routledge is an imprint of the Taylor & Francis Group, an informa business

Library of Congress Cataloging-in-Publication Data

Staten, Greg.
The Avid handbook : advanced techniques, strategies, and survival information
    for Avid editing systems / Greg Staten and Steve Bayes.—5th ed.
        p.   cm.
    Previous ed. cataloged under author Steve Bayes.
    Includes index.
    ISBN 978-0-240-81081-2 (pbk. : alk. paper)
    1. Video tapes—Editing—Data processing. 2. Motion pictures—Editing—Data processing. 3. Avid Xpress. 4. Media composer. I. Bayes, Steve, 1959- II. Bayes, Steve, 1959- Avid handbook. III. Title.
    TR899.B37 2009
    778.59'3—dc22   2008026273

British Library Cataloguing-in-Publication Data
A catalogue record for this book is available from the British Library.

ISBN 13: 978-0-240-81081-2 (pbk)

For Kathleen

CONTENTS

PREFACE

When Steve Bayes approached me about taking over *The Avid Handbook*, I don't think I fully realized what a huge responsibility he had handed me. Now in its fifth edition, this book has been an essential tool and reference for thousands of editors around the world. Taking over a book from another author is always fraught with peril. Because I wasn't going to completely rewrite the book, the result is a merger of our two voices and styles. Fortunately, I discovered that Steve and I have very similar writing voices, so the merger went well.

The world of editing has evolved in a number of ways since the publication of the last edition, most notably with the expectation that editors will be comfortable working in both standard and high definition, which wasn't really on the radar when the last edition came out. Fortunately, much of the existing information in the book was still relevant and I was able to focus my attention on new sections, such as a complete discussion of standard and high-definition video signal, a much deeper discussion of finishing workflows, and color correction. I also took this opportunity to reduce the amount of information covering earlier Avid hardware platforms including the Avid Broadcast Video Board and the Meridien system. There is still some information on Meridien—especially as those systems are still in use in some markets—but the focus of the book is on the newer hardware, DNA- or DX-based hardware.

Media Composer has dropped dramatically in price since the last edition of this book, with a software-only version selling for U.S. $2495, a price that many folks probably never thought they'd see. (I remember that the first Avid system I worked on nearly 16 years ago cost nearly $100,000.) But though the system has dropped in price and evolved in capabilities, the core remains the same as it was all those years ago. Media Composer continues, in my opinion, to have the deepest trim toolset of any system on the market. I feel so strongly about this that Chapter 2, one of the new chapters in the book, is almost entirely focused on trim. Chapter 1, similarly, focuses on methods of editing. These two initial chapters are designed to introduce you to the core power of the system, and I strongly encourage all levels of editors to read through them, especially because they don't just cover the basics but also delve into the deep techniques buried in the system.

You'll also notice that this edition includes sidebars and tips. The sidebars are displayed in gray boxes and contain discussions of topics that are either somewhat peripheral to the main topic being discussed or expand on a topic mentioned in the main body. Tips and notes are also provided and run in the margin to the outside of the main body. These are typically also displayed against a gray background in small boxes. Tips are called out with a thumbtack icon while notes are called out with an exclamation point icon. These are designed to supplement the main body providing additional information or guidance on the topic being discussed.

You will also on occasion notice a flag icon by itself to the outside of the main body. These flags indicates a feature that was added in version 3.0, the latest version of Avid Media Composer as this edition goes to press. If you are running a previous version, the material being discussed in the main body may not be applicable to you.

The Icons Used in this Book

 Web Link — External websites that offer additional resources or information.

 Noteworthy — Learn important "gotchas" or pitfalls that can put your production at risk.

 Technical Tips — How-to's or important advice on how to get the job done.

 Flag — New features added in Avid Media Composer, version 3.0.

This book is intended for overworked editors and assistants who find themselves needing to know more about the system than their limited—or perhaps nonexistent—training has provided them. Though you could certainly read this book from cover to cover, I encourage you to treat it nonlinearly. If it were fiction it would be a collection of short stories rather than a novel. Jump to the section you need to learn more about and dive in! Then put the book proudly back on the shelf for later reading and reference.

If you are a professional whose career is editing on Avid, by all means find the time to take a class or find a good teacher or mentor. Spend time with other experienced editors in your facility or at one of Avid's user groups. There is a whole world of knowledge out there waiting to be explored and experienced.

I'd like to thank a few folks without whom this book would not have been possible. First on this list is Steve Bayes for believing that I could shepherd his baby onward into the future. I'd also like to thank Curtis Poole with Avid's training services. Some of the content in this book—most particularly the chapters on standard-definition and high-definition video signal—I had originally written for training courses while a member of Avid's excellent Training Services group. It is with Curtis's blessing that this material is reproduced here. I'd also like to thank Ashley Kennedy for reviewing some of the content in this book and allowing me to use material from her short documentary, *Common Art*, for some of the visual illustrations. And thanks to Fife Productions for allowing me to use the Nashua Symphony promotional material and shots from their Epic Australia production.

Finally, I have to thank my wife, Kathleen, for putting up with me as I took time away from our family to write this book. I know that there is no way I could have done this book without her support and reassurance. I love you, my wife.

1

ASSEMBLING THE TIMELINE

"Throw up in the morning. Clean up in the afternoon."

—Ray Bradbury

Though there are many approaches to an edit, many years ago a friend showed me the quote above, which is perhaps the best explanation of the editorial process I've ever seen. In other words, get the elements you need into the timeline first. Once they're there you can refine and fine-tune them until you get to the final result. These two phases of the edit are where the storytelling is done and where we'll begin our exploration of Avid Media Composer®. In this chapter we'll look at the techniques and approaches you would use in the rough-assembly phase. In Chapter 2 we'll explore the fine-tuning phase.

Building the Story Framework

When it comes to adding material to a timeline, there are two different approaches you can take. The first is the classical source-to-record process where a clip is loaded into the source monitor, marks are made, a location for the edit is selected in the timeline, and then the desired material is added to the timeline. The second is by selecting a clip or clips in a bin, then dragging them to the desired location in the timeline. Both have their advantages and disadvantages. You could, if you wished, build your entire sequence using only one of these two techniques. But if you really want to master the tools that Avid provides you with, you should become comfortable with both.

Source-to-Record Editing

As this is the classic approach, long-time video editors will probably be most familiar with this method. But if you're more familiar with the drag-and-drop approach to editing, you may find some of these techniques to be a revelation. Even in the rough-assembly stage, the precision available with this source-to-record editing can be a real time-saver.

Rather than discuss the basic workflow for editing from source to record, let's take a look at some techniques you can use to help with your speed and precision. One of the points to remember about Avid is that there are always multiple approaches that can be used to tackle any problem. For that reason I've presented the techniques below, organized into categories rather than by work-flow stages.

Finding the Edit Point

Once you've loaded the desired shot into the Source monitor, the most common approach to finding the edit point is to push Play and then either place a mark or stop when you reach the desired location. You could also just grab the position indicator and drag it right or left, scrubbing the clip until you find the point you want. Finally, you can also use the frame step keys (mapped by default as numeral keys 1, 2, 3, and 4 on the main keyboard) to move forward or backward by either one or ten frames. All of these approaches work, but there are some additional tools available to you that can really help you find the right place for your mark or edit.

Digital Audio Scrub

One disadvantage to finding your point by dragging the position indicator is that you can see the picture, but you can't hear the sound. Digital Audio Scrub is designed to address that limitation. When enabled, you hear individual frames of audio as the position indicator passes over them. To enable Digital Audio Scrub:

- Press the Caps Lock key to turn Digital Audio Scrub on. It remains on until you press the Caps Lock key again to turn it off.
- Hold the Shift key down while scrubbing. Using the Shift key will only activate Digital Audio Scrub while it is depressed.

Digital Audio Scrub is most useful when finding the beginning and ending of distinct sounds, such as the beginning and ending of a sound bite. It is, to be honest, fairly useless when trying to find a point in music or even dialog recorded on location in a noisy environment. Indeed, you'll probably find it to be more annoying than useful in those situations!

Despite this, give it a try. You may find it to be one of the fastest techniques available to quickly hit the beginning and end of a sound bite. But please, for the sake of those nearby and perhaps for your own safety, turn it back off after you've used it to find your mark. There are few things more annoying to others within earshot than a continual blip, blip, blip every time you move to a new position in a source or in the timeline. There is a reason why some editors refer to the Caps Lock key as the "torture key"!

I strongly recommend that you use the Shift key instead of the Caps Lock key when using Digital Audio Scrub. That way it is only on for the brief moment of time that you need it on. Believe me, everyone around you will appreciate it. But there is one "gotcha" to using the Shift key: If you want to use it along with the single-frame step keys (mapped to the 3, 4, ←, and → keys by default), you can't have anything else mapped to the "shifted" state of that key. It is for this very reason that the left and right arrow keys on the default keyboard have the single-frame step commands mapped to each key's normal and shifted state.

J-K-L Scrub

This is quite possibly the most versatile feature in the system. If you aren't already using it then it is time to start! J-K-L Scrub is very powerful because it gives you access to all of the following capabilities in just three keys:

- Play forward or backward at sound speed (i.e., 29.97 frames per second [fps] for NTSC, and so on).
- Shuttle at high-speed forward or backward.
- Scrub at quarter-speed forward or backward.
- Scrub forward or backward by one frame while hearing audio.

Best of all, you can do all of these not only while looking through your footage or your sequence, but also while trimming it. Also, if your deck supports the full Sony command set, you can also use it while shuttling through a tape.

We call it "J-K-L" Scrub because those are the keys the *Play Reverse*, *Pause*, and *Play Forward* commands are mapped to by default. But you can map them to any keys. For example, on my system I have them mapped to D, F, and G on the left half of the keyboard. Regardless of where you map them, the functionality remains the same. Table 1.1 lists how to access the various play modes. (*Note:* If you have remapped these commands, press those keys instead.)

J-K-L shuttling is great because you can dynamically switch on-the-fly between all of the play modes listed in Table 1.1. This means you can roll forward at 2× speed, switch to 1× reverse speed when you roll past the point you want, then play forward and backward at either quarter speed or frame by frame until you

Table 1.1 J-K-L Scrub Operation

Operation	Key Usage
Play forward at sound speed	Press L key
Play reverse at sound speed	Press J key
Pause playback	Press K key
Play forward at faster than sound speed	Press L key twice for 2×, three times for 3×, four times for 5×, five times for 8×*
Play reverse at faster than sound speed	Press J key twice for 2×, three times for 3×, four times for 5×, five times for 8×*
Play forward at quarter speed	Hold K key, then press L key
Play reverse at quarter speed	Hold K key, then press J key
Scrub forward by one frame	Hold K key, then tap and release L key
Scrub backward by one frame	Hold K key, then tap and release J key

*The sound only plays at speeds up to 3×; once you hit 5×, the sound, thankfully, cuts out.

find the exact frame you want. This technique is actually similar, and uses the same default keys, as a linear tape editor used on an edit controller to shuttle through a tape. But shuttling on a computer is far faster than it could ever be on tape, as decks just can't respond as quickly as a digital system.

Soon you will find yourself cooking through the material at double or triple speed while following the script. Surprisingly, you'll be able to understand what people are saying and can work consistently at the higher speed, flying faster through the material and more quickly finding what you're looking for.

One distinct difference between J-K-L Scrub and Digital Audio Scrub is that J-K-L has a more "analog" sound, especially when scrubbing at quarter speed. Long-time editors (those who have been in the business long enough to edit on open-reel decks) often refer to J-K-L as "rocking reels," as the sound really does match what you'd hear if you were manually scrubbing open-reel audiotape with your hands. As a result, J-K-L Scrub is especially useful for hearing inhales and exhales. When heard at quarter speed, a breath has that distinct "Darth Vader" sound that makes it so easy to hear when someone has finished exhaling or inhaling. Once you start using J-K-L you'll wonder how you ever managed to cut without it.

Seeking a Specific Timecode

In some cases you may be working with a producer who has screened the footage and has noted a series of "great lines" or similar points and given them to you in an email. Media

Changing the Way J-K-L Scrub Changes Speed

By default, J-K-L Scrub instantly reverses direction if you switch between forward and reverse play, regardless of the speed you were working at. Many editors prefer this method as it allows them to instantly change direction when they roll past a section they were looking for. But some editors prefer to use J-K-L as much as they would a shuttle knob on a deck. In this case, for example, if you are rolling forward at 3× and turn the knob slightly counterclockwise, the deck slows down slightly, perhaps to 2×, but continues forward. In order to reverse direction, you must roll the knob further counterclockwise through pause and then into reverse.

This approach is known as *speed ratcheting* and you can configure J-K-L Scrub to ratchet if you wish. To use speed ratcheting with J-K-L:

• Hold the Alt/Option key down while pressing the J, K, and L keys.

The following illustration shows how J-K-L ratcheting works.

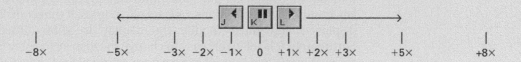

If you prefer this scrubbing style simply map the Alt/Option modifier to the J, K, and L keys on your keyboard. Then the modifier will be applied automatically, and you don't need to add it yourself.

Composer allows you to easily seek, or jump to, any Society of Motion Picture and Television Engineers (SMPTE) timecode that exists in the loaded source clip. Of course, that means that the timecode has to *exist* in that clip; if you seek a point outside the timecode range of a loaded clip (e.g., if the clip has timecode from 06:25:05:01 to 06:27:06:25 and you ask it to seek to 06:27:15:00) the system will merely beep at you. To seek to a timecode:

• Load the desired clip into the Source monitor and ensure that the Source monitor is active. (If you aren't sure, click on the Source monitor to activate it.)

Using the keyboard's numeric keypad, type the desired timecode and press Enter to seek that timecode. (If you are working on a laptop you cannot use the number keys above the letters to enter timecode. Instead you must use the Fn key to enter the numbers using the alphanumeric keys. See your laptop's manual for more information on using the Fn numeric keyboard.)

Save Keystrokes While Entering Timecode

You do not have to enter the colons (or semicolons if using drop-frame timecode); the system will add them for you automatically. In addition, you only have to enter the portion of the timecode that is unique from the timecode of the frame you parked on. This means that if you are parked on a frame with timecode 01:02:09:08 and you wish to seek to frame 01:02:25:00, you only have to type "2500" and press Enter. The system will use the hour and minute from the current frame. In addition, you can press the period (.) key on the numeric keypad to enter two consecutive zeros. This means, for example, that you can seek timecode 01:00:00:00 by typing "1…" on the numeric keypad.

There is a potential "gotcha" that could prevent you from seeking any timecode, even that which exists in a clip. Media Composer seeks based on the time format displayed above the Source or Record monitor. Depending on how your system is configured, you may not be displaying timecode but instead frame numbers, film key numbers, clip durations, or even clip names. If timecode is not displayed you must modify the display to show timecode. Fortunately, it is relatively easy to change the display to show timecode. (*Note:* If two lines of information are displayed, only the top one must be displaying timecode. The lower line can be used to display any other information desired.)

To set the source information display to timecode:

1. Click on the information display you wish to modify to show the menu. The top of the menu allows you to set the type of information you wish to use. Below that are various types of data that are valid for the type of data currently selected.

2. Move the cursor to the Source option at the top of the menu and a submenu will display listing all the tracks in the loaded source clip.

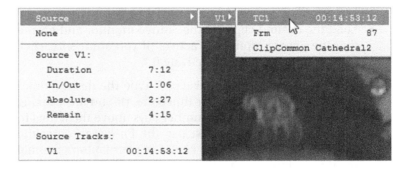

3. Move the cursor again to one of the source tracks and another submenu will display that lists the types of information available. For video projects you will see TC1 (timecode), Frm (frame count), and Clip (clip name). If you are in a film project you will see additional types of information including key number, ink number, and so on.

4. Select "TC1" from the menu to change the information display to timecode.

Searching for Timecode across Clips in a Bin

During capture, often a camera tape is broken down into many, perhaps dozens, of master clips that contain key sections, or selects, from the tape. But, perhaps the producer who is providing his or her list of timecode points may have been watching a copy of the tape and hands you a list of timecode references that don't directly correspond to the master clips you created. Fortunately, Media Composer allows you to search across an entire bin to find a clip that contains a specific clip. This is accomplished using the Sift command. We'll discuss Sift in detail in Chapter 3, but for now let's look at how we would use it to find the master clip containing the timecode we need to seek to.

1. Open or select the bin containing the clips from the tape the producer logged.
2. Choose a bin view that contains at least the Start or End timecode column. (The built-in Statistics view contains both of these columns. If you are on Media Composer 3.0 or later you can also choose the Capture view.)
3. Choose "Bin > Custom Sift…" to open the Sift dialog.

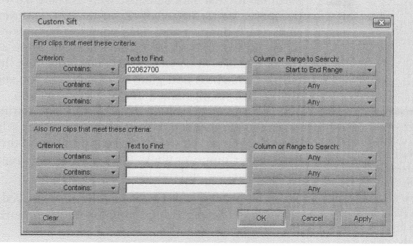

4. From the top line enter the *entire* timecode you wish to find. It is not necessary to enter the colons or semicolons, but you must enter the hours, minutes, seconds, and frames.
5. Click either Apply or OK to perform the sift.

The clip (or clips) that contain that timecode will be displayed in the bin. All other clips will be hidden. To show them again, choose "Bin > Show Unsifted."

You can also seek by timecode in the sequence as long as the sequence timecode (Mas or Mas TC) is displayed in the highest information view above the Record monitor.

Making Your Marks

To make an edit you need to set marks and a point of sync. I'm not going to go into the hows and whys of three-point editing, but there are some details regarding the way Media Composer either lets you mark or responds to those marks that are worth discussing. Perhaps some of these are new to you!

Checking Your Duration

One of the handiest features in the Composer window is Center Duration. This useful feature provides you with a single location where the marked (or unmarked) duration of the active monitor is always displayed. If you are migrating from Xpress Pro® this feature will be new to you, but I'm often surprised at the number of Media Composer editors who don't know about it. How is that so? Well, the feature has always, for some reason, been disabled by default. Fortunately, that has changed with version 3.0 of both Media Composer and Symphony®. If you create a new user in these versions you'll discover that your new user has this option, among others turned on by default. You'll also find a new set of interface colors installed when you create a new user. (Yes, that is right, the purple highlight is now gone by default. You can still access it via an interface setting named "Classic" if you miss it.)

If you don't have this enabled in your user setting, you can do so via the Window tab in the Composer setting. When enabled it displays the marked (or unmarked) duration of the active monitor. Media Composer obeys the following rules regarding marks, or lack of marks, and the displayed duration:

- *In and Out mark:* Marked duration displayed.
- *In or Out mark only:* Duration between mark and position indicator.
- *No marks:* Duration from position indicator to end of active clip or sequence.

By default, Center Duration displays the duration in SMPTE timecode. If you click on the time display you can switch it between timecode and frame count. If you are in a film project, you can also switch it to feet and frames.

What Is It with Media Composer Version Numbers?

You might wonder why, if Media Composer is 20 years old, the latest version, as of this printing, is called 3.0. Prior to the release of the Adrenaline® hardware in 2003, Media Composer had reached version 12.0. Whether it was a superstition for the number 13, a fear that the version number had gotten too high, or some other reason understood only by those who made the decision, Media Composer was reset to version 1.0 when Media Composer Adrenaline was released. If the numbering had continued we would have seen version 15.0 of Media Composer released in 2008. (Similarly, Avid Symphony had reached version 5.0 before it was reset to version 1.0 with the release of Avid Symphony Nitris® in 2005.)

Note: In 2008, Media Composer and Symphony were sychronized at version 3.0, which simplifies the version numbers somewhat.

Three-Point Editing

The most fundamental method of editing is, of course, the three-point edit. Remember that the In and Out marks indicating edit duration can be made in either the Source or the Record monitor. And, in the absence of the solo In mark, the Avid system will use the location of the position indicator.

Back-Timing an Edit

A three-point edit doesn't have to be two In marks and one Out mark. If you want to use the Out point as the sync reference, then use two Outs and one In!

"Mark and Park" Editing

Lightworks®, another digital film editing system popular in the early 1990s, was configured so that an editor never had to mark an Out if he or she didn't want to. The editor could instead mark the In and then park on the frame he or she wanted to cut out on. You could argue that all this does is save a single keystroke, but those single keystrokes can add up over time.

Media Composer also provides this functionality under a feature called "Single Mark Editing." You can enable this under the Edit tab of the Composer setting. Once set, all you need to do is mark an In (or an Out for a back-timed edit), move your position indicator to the frame you want for the other side of your edit, and then either Splice or Overwrite the footage into the timeline. (*Note:* If this feature is not enabled, marking just an In point will edit in from the mark to the end of the clip.) You might want to give this feature a try; I know a lot of editors, especially feature film editors, who swear by it.

Changing Your Marks

If you want to change the position of your In or Out mark, you can simply move the position indicator to a new position and remark. But you can also drag an existing mark to a new location. Simply hold the Alt/Option key down, click on the mark below either the Source or Record monitor, and drag it to the desired position. (You must drag from the timebar beneath one of the monitors; you cannot drag marks on the timeline.)

Previewing Your Edit

Media Composer contains a nice feature known as Phantom Marks that allows you to see the "fourth" mark in a three-point edit. This feature is especially useful when doing an Overwrite or back-timed edit as you can see the frame duration that will be affected prior to performing the edit. If you wish to turn these marks on, you can do so via the Edit tab of the Composer setting.

These marks are incredibly useful, but unfortunately the color used for these marks is just slightly bluer than the regular gray marks. On a high-resolution display or any display after eight hours of editing, I can promise you that you'll go cross-eyed trying to discern which mark is real and which is a phantom.

Fortunately, *all* of the functionalities for Phantom Marks are available when this option is disabled! This feature merely turns on the *display* of these marks. If you have marked for a three-point edit, all of the following functionalities are available for the side with only one mark:

- *Go to In/Go to Out:* If you've marked an In, pressing Go to Out will jump to the last frame that will be edited in or over, based on the marked duration in the other monitor. And if you've marked an Out, pressing Go to In will jump to the first frame that will be edited in or over. As you can imagine, this is extremely useful for back-timed edits!
- *Play In to Out:* This plays the duration marked on the other side. It will either play *from* your marked In or *to* your marked Out.
- *Play to Out:* This plays from your marked In to the last frame that will be edited in or over. This command is not very useful for back-timed edits, but is the perfect command if you are parked on your In point.

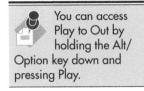

You can access Play to Out by holding the Alt/Option key down and pressing Play.

Be careful, though! Remember that if you don't have any marks, Media Composer uses the position indicator as the In point. That means if you don't have an In or Out marked when you use either Go to Out, Play In to Out, or Play to Out the position indicator will move to a new location, which will then become your new In point for the edit. It might seem odd that it behaves that way, but it is utterly logical when you think about it. (And remember that computers are nothing if not logical.)

Marking a Timecode Offset

If you wish to mark a specific duration you can mark either an In or an Out then use the numeric keyboard to move the position indicator to a specific number of frames then add the other mark. This is easily accomplished by adding a "+" either at the beginning or end of the offset number, then hitting Enter. But remember, Media Composer uses a film-based counting method that includes every frame within the marks. This means you'll usually want to move one frame less than the duration desired. See the next sidebar, "Counting Frames the Avid Way," for more information on why Avid works this way.

Most of the time you probably want to enter a timecode duration such as three seconds (3:00), but sometimes you want to move a specific number of frames, not seconds. If you type three digits in Avid, it automatically assumes you are entering timecode and assigns the first digit to seconds and the third (and fourth) to frames. If you really want 300 frames instead of three seconds, simply type a lowercase "f" after you've typed your numbers but before you hit Enter. The "f" tells the Avid to count in frames and you'll notice that it immediately recalculates the timecode using the number of frames you specified.

Counting Frames the Avid Way

Avid editing systems work from the film model that insists that every frame exists and is important to duration calculations. In contrast, the linear tape model allows you to have an Out point and an In point on the same frame of timecode. How can two shots use the same frame on the master tape? Don't ask—in linear video editing they just do, and video editors who grew up editing linearly are occasionally confused when the Avid editing system counts every frame discretely. Mark an In and Out point on the same frame in Avid. What's the duration? One frame. And if you are parked on a frame and mark it as the In point, then tell the system to "Go 15 frames from here" by typing "+ :15," then mark Out, you will have a duration of 16 frames. You told the system, "Take this first frame and 15 more," which makes perfect sense to a film editor and seems suspiciously like a bug to a video editor.

This is why if you want to mark, for example, a five-second duration, you must subtract one frame from your offset duration. You can either do that, as most Avid editors do, by typing "+4:29" (for NTSC 30 fps timings; use "+24" instead for PAL timings) instead of "+5:00"; or, type the full duration then back up a frame before marking your Out point. Personally, I find "doing the math" in my head is quicker than moving back a frame after the fact.

By the way, when you enter a frame offset via the numeric keypad, Media Composer automatically stores that offset. If you need to use it again simply hit the Enter key on the numeric keypad without entering any numbers and the system will automatically move again by that amount. This is an especially useful tool when marking out beats either for edits or for effect keyframes.

Marking a Segment in the Timeline

You should already know that, for a selected track, you can park in the middle of a segment and press Mark In-to-Out to mark the entire segment. But if you have multiple tracks selected, doing so will mark the nearest points in either direction where the tracks share a common edit point. Sometimes that means the entire sequence is marked. If you only want to mark the duration of the shortest segment across all tracks, simply hold the Alt/Option key down when pressing Mark In-to-Out.

Snapping to Edit in the Timeline

Sometimes you need to mark several segments. In this case, you can use the Ctrl/Command key to snap to either the head or tail of an edit point. This makes it easy to quickly mark a series of shots for quick replacement or removal.

- Press Ctrl/Command and click the mouse button to snap to the head of the nearest edit. You're now properly positioned to mark an In.
- Press Ctrl+Alt/Command+Option and click the mouse button to snap to the tail of the nearest edit. You're now properly positioned to mark an Out.

You can also use the FF (fast forward) and REW (rewind) keys to move from edit to edit. By default they snap to the heads of edits and, just as is the case with Mark In-to-Out, only move to common edit points if multiple tracks are selected. If you want to jump to every edit point regardless of track selection, just hold the Alt/Option key down, just as you did with Mark In-to-Out. Or, you can go even further if desired. The FF/REW tab in the Composer setting allows you to reconfigure these two commands. You can force them to move to every edit on each track (by choosing Ignore Track Selectors) and even instruct them to stop at tail frames of an edit and/or locators. I rarely set them to stop at tail frames, preferring to use the Ctrl+Alt/Command+Option click for that; however, setting them to stop at locators can be extremely useful, especially when reviewing a screening with your client or producer. Indeed, I have a special Composer setting I switch to in these situations where the FF/REW keys are set to *only* jump to locators, allowing me to quickly move from comment point to comment point.

Editing to the Timeline

Certainly the most common edits are Splice, which always adds your marked frames to the sequences, and Overwrite, which generally replaces footage that already exists in the sequence. But there are two other edits that are extremely powerful.

Sync Point Overwrites

The Sync Point Overwrites edit is a special configuration of the Overwrite edit that changes the way the two Source and Record sides are synchronized. Remember that normally Avid uses either two In points or, in the case of a back-timed edit, two Out points as the point of synchronization. A Sync Point Overwrite only uses an In and an Out to specify the *duration* of an edit. The blue position indicators are used as the points of sync.

Sync Point Overwrites *cannot* be three-point edits for this very reason. Indeed, you must have only one In and one Out point or the edit will fail. The marks can be either both in the Source or Record or, less commonly, one in the Source and the other in the Record.

This type of edit is especially useful for cutaways and inserts because often the point of sync is somewhere in the middle rather than the beginning or end of the edit. For example, if I wanted to cut to an insert of a glass being dropped on the floor, the point of sync is likely to be the sound of the glass when it hits the floor. The In and Out points of the edit are used just to establish the timing around the drop.

As mentioned previously, accessing this type of edit requires you to change the Overwrite edit's configuration. You can do so two different ways:

- Select "Sync Point Editing (Overwrites)" from the Edit tab of the Composer setting.
- Select the Composer window then choose "Special > Sync Point Editing." You can also right-click on the Composer monitor and choose "Sync Point Editing."

Regardless of the technique used, the Overwrite button's icon will change, and an orange dot will be added in the middle of the arrow.

To use a Sync Point Overwrite:

1. Mark the desired duration for your edit in either the Source or Record monitor.
2. Move the position indicator to the appropriate sync point for each side of the edit.
3. Press Overwrite to perform the edit.

Because this edit replaces the normal Overwrite edit, you need to make sure to turn it off once when you are ready to return to

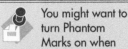 You might want to turn Phantom Marks on when using Sync Point Overwrites as it will make it easy to see the duration you'll edit in without forcing you to reset your point of sync.

normal Overwrite editing. Some editors choose to leave it on, but have to remember to properly place their position indicators and make sure to only use two marks.

Replace Edit

Perhaps even more powerful than the Sync Point Overwrites is the Replace edit. This edit does not require *any* marks in either the Source or Record side. (Indeed, it does not allow *any* marks in the Source.) Instead it uses existing edit points for a segment in the timeline as the duration and the two position indicators as the point of sync. In this respect it is extremely similar to the Sync Point Overwrites with the exception that the duration is always specified in the sequence rather than the source.

The Replace edit button looks similar to the Splice and Overwrite edit buttons, but has a blue arrow instead of the yellow and red arrows used for Splice and Overwrite, respectively. On early versions of Media Composer, the Replace edit lived between the Splice and Overwrite buttons at the bottom of the Composer window. On modern Avids, it lives in the command Fast menu that resides between the Splice and Overwrite edit buttons. You can, of course, map it anywhere you wish, including putting it back in its old location. (I personally map it to Shift+B so it lives on the same key as the Overwrite edit.)

The most common usage is to replace an entire segment in the timeline with another shot or take. When I'm doing an online finish of a show someone else cut, I find myself particularly using the Replace edit if the legal review of the program requires the replacement of some of the B-roll or interview material used in the edit. Interview material may require some trimming and careful adjustments to ensure that I don't change the program duration (see Chapter 2), but the Replace edit is certainly the safest way to replace footage in a locked-to-time program.

To perform a Replace edit:

1. Place the Timeline's position indicator inside the segment you wish to replace.
2. Place the Source monitor position indicator at a point of sync to match the Timeline's position indicator.
3. Press the Replace edit button.

What isn't well known about the Replace edit, though, is that it *can* use marks instead of a segment for its duration. The only limitation is that these marks can only reside in the timeline, not in the source. When used with marks, the Replace edit behaves identically to the Sync Point Overwrites. That means that if the timing of your insert or cutaway is based on the sequence duration, you can use the Replace edit instead of switching the configuration of the Overwrite edit.

Even less well known is that you can actually use a segment's duration *but* set the point of sync to a location *outside* of the segment. This is most frequently used to replace a split edit. If you've split an interview so the sound cuts before the picture, you'll likely want to use the start of the sound bite as your point of sync. Since the head of the audio segment is the start of sync, replacing the audio is easy. But the picture edit start is delayed by the split edit. How do you replace it so it stays in sync with the sound?

Easy! Place the position indicators appropriately so they are lined up in sync, then use one of the segment mode buttons to select the desired segments as shown in the following illustration.

When you perform the Overwrite edit, the entire split edit (both video and audio) will be replaced with the new synced material. If the old and new interview sound bites have a different duration, you'll just need to trim the tail. In this way, the Replace edit allows you to keep the split you already positioned, instead of unsplitting, replacing the edit, then resplitting! This technique has saved me lots of time in the online—and even color-grade—stage.

Editing from a Pop-Up Monitor

This next technique is probably familiar to those who have edited with Xpress Pro or NewsCutter® and is tailor-made for cutting back and forth between two or more clips. If you hold the Alt/Option key down while double-clicking on a clip in the bin (instead of loading into the Source monitor) the clip will load into a pop-up monitor. You mark an edit from a pop-up using the same keyboard shortcuts as you would with the Source monitor. If you prefer to use the onscreen buttons instead, you must use the buttons at the bottom of the pop-up monitor instead of the buttons in the Composer window. If you're cutting back and forth between two clips, you will find this approach quicker—and certainly less carpal tunnel–inducing—than switching between the two clips in the Source monitor.

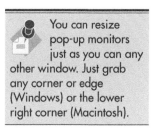
You can resize pop-up monitors just as you can any other window. Just grab any corner or edge (Windows) or the lower right corner (Macintosh).

Editing from the Bin

Rather than loading material into the Source monitor, making your marks, and editing that footage into a sequence, many editors today prefer a more interactive approach where they simply grab clips from a bin, drag them to the timeline, then drop them at a location of their choosing. If an editor started working with another nonlinear editor such as Final Cut Pro® or Premiere® this is likely the technique they first learned.

Many long-time Avid editors would argue that the classic source-to-record approach is the best, but dragging clips from a bin can be a very fast and efficient approach, especially if you have either logged selects from tape or subclipped out the material you plan on using. Personally, I find this technique especially useful when I'm quickly putting a B-roll section together, or when I'm dropping audio sound effects, music, and stings into my sequence.

Basic Drag Techniques

Though the basics of dragging a clip from a bin to the timeline are certainly obvious, there are some subtleties to the way it currently works on Avid.

Choosing the Type of Edit (Splice versus Overwrite)

When you drag a clip to the bin, Media Composer assumes that you add new material to your sequence and automatically selects the Splice segment editing. But that isn't always what you want. You may, instead, want to perform an Overwrite. You can do so, but you must tell the system that is what you want to do before you begin your drag.

To switch to Overwrite drag-and-drop editing:

- Select the Lift/Overwrite (red arrow) segment mode by clicking on the red arrow at the bottom of the timeline.

Map the Lift/ Overwrite segment mode button to a key on the keyboard so it is quickly accessible.

As long as the Lift/Overwrite segment mode is enabled, every clip dragged to the timeline will be overwritten. I typically use this approach when I'm adding audio elements to the timeline. Splicing them in will push later material on the track down in time and break its sync. When you want to switch back to the Splice segment editing, simply turn the Lift/Overwrite segment mode off or select the Extract/Splice (yellow arrow) segment mode.

Dropping the Clip Precisely

When you're dragging a clip to the timeline, you'll often want to drop it in a specific location, such as between two existing edits. If you're dropping in a cutaway you might want to drop it so that the cutaway begins after a specific frame of a shot already in your timeline. Media Composer has tools that help you accurately perform both types of precise positioning.

Modifiers and Dragging from a Bin

Typically, when you use a modifier to affect a mouse click or drag, you hold the modifier down *before* you press the mouse button. For drag-and-drop editing, though, you must first click the mouse and *then* add the modifier. Why? Because these modifiers are also used to change a bin selection or bring up information windows.

For example, if you hold Ctrl+Alt/Command+Option down and click on an item in the bin, an Info window appears that displays a set of metadata about the clip including its tape, tracks, video resolution, duration, and so on.

Make sure, though, that you hold the modifier down until you've released the mouse. Otherwise the system will think you changed your mind and not apply the modifier to the drop.

Name	Eric - what separates us
Tape	006NSP50706
FPS	29.97
Tracks	V1 A1-2
Video	DNxHD 145
Audio SR	48000
Drive	WD Passport (E:)
Start	10;03;05;08
End	10;03;24;27
Duration	19;19
Project	Classical Music
Audio Bit Depth	16

The Avid offers a set of modifiers that can be used to tell the system to snap to different places in the timeline. The most commonly used of these is the Ctrl/Command key, which restricts the drag-and-drop in a number of very useful ways:

- Snap to the head of an existing edit on any track in the timeline.
- Snap to the head of the position indicator.
- Snap to an In point.
- Specify the duration used in the source clip.

What is that last one, you say? Specify duration? Absolutely! If you use In and Out marks to set a duration in the timeline, then drag a clip while holding down the Ctrl/Command key, the clip is not only snapped to the In point, but the Out point defines the duration you will edit in from your source clip. As you are likely dragging in clips without a defined duration (other than the complete duration of the source clip), this technique makes it easy to drop in only the duration you need.

This works for both Splice and Overwrite drags, but keep in mind that if you Splice drag, there will be a new edit at the In point and all of the existing footage at that point will push down to make room for the new clip.

Another modifier worth knowing is Alt/Option. When you drag a clip into the timeline your ability to accurately drop it at a specific frame is limited by the pixel resolution of your screen and your timeline zoom level. If you're really zoomed out on a long sequence, a single pixel move may actually move the clip several seconds forward or backward. If you hold the Alt/Option key down, though, you will *always* move a frame at a time forward or backward. Indeed, as you drag back and forth while holding these modifiers down, you may see your cursor move further than the actual clip moves. This is because it is moving a distance less than a single pixel, which your cursor cannot do.

A third modifier worth knowing is really useful when doing an Overwrite drag. If you hold the Ctrl+Alt/Command+Option modifiers down, an Overwrite drag will snap to the tail of an edit, allowing you to back-time the clip into the timeline. Unfortunately, this modifier does not work for a Splice drag. Instead of back-timing, the Splice drag will always occur at the position of the head of the shot you're dragging in.

Similar to the Ctrl/Command modifier, the Ctrl+Alt/Command+Option drag will snap to tails of edits, the tail of the position indicator or a mark. It does *not*, however, snap to a duration—that capability is reserved for the Ctrl/Command modifier.

There's a fourth modifier but it really isn't applicable for a drag-and-drop edit. We'll discuss it when we explore Segment mode later in this chapter.

The Drag-and-Drop Viewer

When you drag a clip from a bin into the timeline, the Composer window switches to a four-monitor view. This view allows you to see the frames from the sequence that will be just before and just after the clip you are dropping.

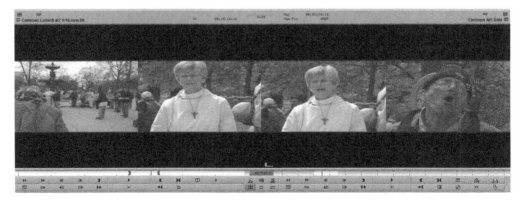

This display can be extremely useful in conjunction with the Ctr+Alt/Command+Option modifiers to precisely align the clip you're dropping. It can also be helpful with snap-to-head or snap-to-tail drops as well because it lets you confirm that you are dropping the clip between the right two shots.

On older, slower systems this display can sometimes take a moment to display. If you're doing a lot of drag-and-drop edits, that momentary delay can get extremely annoying. For that reason, there is a way to disable the four-monitor view:

- Deselect "Show Four-Frame Display" from the Display tab of the Timeline setting.

Dragging Sync Material to the Timeline

When you drag a video-only or audio-only clip to the timeline you can specify which track (or tracks) the clip will be dropped onto. In previous versions of Media Composer, you could not do that if you were dropping a clip containing both video and audio tracks. Fortunately, that has changed in version 3.0. Now when you drag a video/audio clip to the timeline, you can drag up or down to specify the track the video will be dropped onto. Unfortunately, that clip's audio tracks will only drop onto to the audio tracks that match in number (A1 to A1, and so on). Ultimately, it would be great to be able to specify the track for either video or audio, and hopefully we'll see that in a future release.

Dragging Multiple Clips to the Timeline

If you select multiple clips and drag them as a group to the timeline, the drop behavior is quite different from that of a single clip. Instead of letting you drag them to any location in the timeline, the selected clips are always edited in at the head of an In point, or, if there are no In points in the sequence, the head of the position indicator.

In addition, the clips are always ordered as they appear in the bin. If you are displaying the bin in Brief, Text, or Script view, that means the clips are edited in order from top to bottom. If you are displaying the bin in Frame view, they are edited in "reading" order (left to right, top to bottom). Frame view, therefore, is probably the best view to use for this type of editing, as it lets you quickly rearrange your clips the way you want them.

Dragging to the Record Monitor

In addition to dragging to the timeline, you can also drag to the record monitor. As with dragging multiple clips directly to the timeline, clips dragged to the record monitor are always edited in at either the In point or, in the absence of an In, the position indicator. You use modifier keys to specify whether the edit will be a Splice or an Overwrite:

- Hold the Alt/Option key down to Splice the clip(s) to the timeline.
- Hold the Shift key down to Overwrite the clip(s) to the timeline.

Marking in the Bin

By default, when you drag a clip from the bin to the timeline, the entire clip is edited into your sequence. Though this works well for short clips or premarked subclips, the reality is that you usually want to edit in just a portion of the clip into your sequence. This is easily accomplished using either the bin's Frame or Script view. When in either of these views, you can select a clip and then use the keyboard to:

- Rewind to the first frame (the Home key by default).
- Fast forward to the last frame (the End key by default).
- Use J-K-L to scrub through the clip.
- Play and stop the clip (the 5 key or space bar by default).

And while the clip is playing you can:

- Mark an In point (the I or E key by default).
- Mark an Out point (the O or R key by default).

Note that you *must* be playing when you mark an In or an Out. This is to prevent you from accidentally changing a mark when you might, for example, be trying to rename the clip.

Editing from the Bin via the Keyboard

After marking a clip in the bin, you can certainly use any of the previously discussed dragging techniques to edit the clip into your sequence, but you can also use the keyboard to edit the clip into your sequence. This option is disabled by default, but can quickly be enabled via the bin settings. To enable Edit from the bin:

- Select "Enable Edit from Bin (Splice, Overwrite)" from the bin settings.

You can quickly open a window/tool's setting dialog by selecting the window or tool and pressing Ctrl/Command+=.

Once you've enabled this option, simply use the keyboard Splice or Overwrite buttons to edit the selected clip into your sequence. You can even edit multiple clips in simultaneously. If more than one clip is selected, they are edited into the sequence in "reading" order, left to right from top to bottom.

Taking this one step further, you can use the arrow keys on the keyboard to move between clips in the bin. Doing so makes it easy to move between clips, play and mark a duration, edit them into your sequence, then move to another clip and continue editing. This technique is especially useful if you wish to build a quick montage from B-roll footage. I find it especially useful when editing an interview or single-camera scene, as using the arrow keys to move back and forth between clips is far faster than switching between two or more clips in the Source monitor.

Disabling and Enabling Tracks from the Bin

Sometimes you only want to use some of the clip's tracks when you edit from the bin. Though you can certainly load them, one at a time, into the Source monitor and disable or enable the desired tracks, you can also do this from Frame view in the bin. This technique can also be used to quickly see which tracks are available in a clip. To disable or enable a clip's tracks from the bin:

1. If necessary, switch to Frame view.
2. Press Alt/Option+Click on the *name* of the clip in the bin. A pop-up menu will appear showing all of the tracks in the clip with checkmarks at the head of those tracks that are currently enabled.
3. With the mouse held down, select a track and release the mouse button to toggle the track off or on.
4. Repeat steps two and three for each additional track you wish to affect.

Auto-Enable Source Tracks

By default, when you load a clip into the Source monitor, all tracks are automatically enabled, even if you previously disabled them. This is because a feature named *Auto-Enable Source Tracks* is enabled by default on most versions of Media Composer. I usually disable this option as I prefer for tracks I've disabled to *stay* disabled, especially because I often use bin editing in the early stages of an edit. This option is located in the Edit tab of the Composer setting.

Rearranging Your Edits

After you've assembled a good portion of your sequence, you may want to rearrange some of the clips you've used. This is easily accomplished using the two segment mode buttons at the bottom of the screen. Though you probably know some of segment mode's basic functionality, you may not know some of the more obscure and newer techniques.

Select Segments via Timeline Lasso

You can quickly select a segment or, more importantly, a group of segments (or clips) by holding the mouse button down and

If you have a lot of tracks it may be difficult or impossible to select above or below the actual tracks in your sequence. In this case, simply hold the Alt/Option key down when you begin your selection. This modifier instructs Media Composer to begin the lasso selection from anywhere in your timeline. Note, however, that you must begin your selection outside any segment you wish to select.

dragging, from left to right, a lasso around them. Any adjacent set of segments can be quickly selected. You must, however, begin your lasso selection either above or below the tracks in your sequence. That is because clicking within a track instructs the Avid system to scrub the position bar. When you release the mouse any segments you've completely enclosed will be selected.

By default, when you lasso a set of segments using the lasso, the yellow (Extract/Splice) segment mode is selected. If you'd rather use the red (Lift/Overwrite) mode, simply click on the red segment mode button and you'll switch modes without losing your selection. Alternatively, you can click on the red segment mode button before your selection, though doing so does prevent you from using the Alt/Option modifier to select from within the sequence's tracks.

If you'd like to add additional segments to your selection, simply hold the Shift key down and click on them. Keep in mind, though, that you must select *adjacent* segments in any track or you will not be able to move them. This means that if there is a section of filler between two segments you must also select the filler. (I know, the concept of selecting the "empty space" between two shots may seem a bit odd, but that's how it works in the Avid world. Empty space is actually a physical thing in the Avid timeline. There are advantages to this but, as you can see, also some disadvantages.)

Moving Segments

Moving multiple segments in the timeline changed in some very significant ways in version 3.0. Because of that, let's look at how segments move in previous releases first and then how they move in version 3.0.

Moving Segments Prior to Version 3.0

In segment mode you can select multiple segments anywhere in the timeline, but you can only move them if the segments are *directly adjacent* (contiguous) on a single video or audio track. Groups of segments can be moved horizontally—as long as they obey the contiguous rule on each track—but only audio segments can be moved vertically. You can also delete selected segments en masse regardless of their location in the timeline by simply pressing Delete on the keyboard.

Moving Segments in Version 3.0

Significant changes were made to segment moves in the latest release. Many of the movement restrictions were eliminated in this version and you are now able to select and move segments horizontally *and/or* vertically across multiple video *and* audio tracks. This means you can now select a group of composited segments and move them both horizontally (in time) and vertically (in track.)

Previously, moving a group of composited segments was a very time-consuming process (or required use of the Avid clipboard).

Note, however, that if you select multiple segments on any given track, the selection must be contiguous, and this includes filler. Make sure that you select the filler between two or more shots so that the empty space is moved as well. Hopefully, this step can be eliminated in future versions, but for now you must select it or the system will refuse to let you move the segments.

In addition, if you have selected both video and audio segments, you can move either the video or the audio segments vertically while moving everything selected horizontally. To do so, simply begin your move by clicking and dragging on the type of segment you want to move vertically. For example, if you want to move the video segments vertically, simply click on one of the video segments to begin your move.

Restricting Segment Movement

If you wish to precisely align the segments you are moving, you can hold a keyboard modifier down. The modifiers listed in Table 1.2 are available.

Table 1.2 Segment Mode Modifiers

Modifier	Restriction
Ctrl/Command	Snap to the head of an existing segment, an In/Out mark, or the position indicator
Ctrl+Alt/Command+Option	Snap to the tail of an existing segment, an In/Out mark, or the position indicator
Ctrl+Shift/Ctrl	Restrict to vertical-only movement
Alt/Option	Force frame-by-frame movement; regardless of how zoomed out your timeline is you will always move frame-by-frame

To ensure the modifier is applied, always release the mouse button before the modifier.

In addition to the modifiers, you can force the system to always snap to the head of an existing segment, mark, or the position indicator by enabling "Default Snap-to Edit" from the Edit tab of the Timeline setting. This was Xpress Pro's default configuration, and may be the preferred method of operation if you're migrating from Xpress Pro.

Cutting Down Your Sequence

While you're still in the rough-edit mode you'll probably need to cut out portions of the sequence. The most common commands are Extract and Lift (which are complements, respectively, of Splice and Overwrite), but there are other ways to quickly remove material. For example, as just mentioned earlier, segment mode can be used to remove multiple shots at once and makes a terrific technique to quickly blow away whole sections of the timeline. And when used in conjunction with the Add Edit command, you can also use it to remove portions of a segment instead of an entire segment.

What Happened to the Weightlifter Guy?

Something that you'll likely quickly notice in version 3.0 is that the Lift and Overwrite icons have been changed and the weightlifter guy has been replaced with an icon using an up arrow. Scandalous, you might say! How could Avid possibly get rid of that goofy icon that we've all come to love? Well, considering I was one of those who was a party to his removal, I'll tell you.

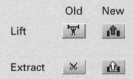

One of the challenges with quirky/idiosyncratic icons and features is that they are difficult to discern by new users or those who "grew up" with a different program. Though we would never want to homogenize the system so that every editing program behaved identically, there are certain assumptions that users make, simply based on how most programs (both editorial and noneditorial) function today. The old lift (weightlifter) and extract (scissors) icons created problems for new users. The lift icon's function wasn't obvious, and, more importantly, the scissors icon actually implied a different function. (Scissors typically are used to indicate a Cut command—as part of the Cut/Paste paradigm—which is a very different operation from Extract.) In addition, the key point that the Lift and Extract commands were complements to the Overwrite and Splice commands, respectively, wasn't clear from their icons.

We did some focus testing and came up with a new set of icons with upward arrows that better informed the user of the function and, by using yellow and red icons, their relationship to Splice and Overwrite. Naturally, there were some long-standing users who mourned the loss of the weightlifter—though no one mourned the loss of the scissors—and one even wrote a very funny ode to the weightlifter guy (which you can find with a bit of Googling or even on my Avid blog, community.avid.com/blogs/editors/). Note that a few other icons changed as well—the "loop" commands in particular, again with the intent of making the icons more obvious to inform users of their functions. I think it is safe to assume that other icons will change in future releases.

Top

Tail

Top and Tail

These two commands first made their appearance in NewsCutter and migrated to Media Composer several years ago.

They are designed to help you quickly cut down existing shots in the timeline by quickly removing either the head or the tail from the segment you are parked on. Interestingly, these two commands are actually built-in macros. Top is the equivalent of pressing Mark Clip, Mark Out, and Extract, while tail is the equivalent of pressing Mark Clip, Mark In, and Extract. These two sets of commands, called TRX/TEX or TIX/TOX by old-school Avid news editors, are such essential quick cut-down commands that they were rolled in to single commands.

Top and Tail aren't mapped to your keyboard by default, but they live on the Edit tab of the Command Palette. I mapped them to Shift+Q and Shift+W on my keyboard.

These two commands make it extremely easy to slice off the unnecessary head and tail of clips you dragged into the timeline or otherwise edited in long. To remove the top of the clip simply park on the last frame you want to remove and press "Top." It will then remove from the first frame to the frame you are parked on. If you wish to remove the tail of a clip simply park on the first frame you want to remove and press "Tail."

Be careful, though, as the first command issued by either of these is "Mark Clip," so you need to make sure that you have not selected tracks that do *not* have common edit points with the track containing the shot you're cutting down. Because the Mark Clip command marks a common duration across all active tracks, you can easily remove more material than you intended—up to and including every frame in the sequence prior to or after your position indicator!

Take a look at the following illustration. Because of the track selection, applying the Top command would remove everything in the timeline preceding the position indicator. To ensure that doesn't happen, all tracks but V1 must be deactivated.

Extract and Lift and the Clipboard

Have you ever wondered what happens to the footage that you either Extract or Lift out of the timeline? When you perform either function, the removed footage is loaded onto the Avid clipboard. You can therefore see and, if desired, subclip or re-edit the footage by choosing "Clipboard Contents" from the Source monitor menu.

Since this capability is unquestionably handy when moving large chunks of material from one part of the sequence to another, you can eliminate the entire Clipboard Contents step by simply holding the Alt/Option key down when you do either an Extract or a Lift. That modifier instructs the Avid editing system to automatically load the removed material into the Source monitor, ready for use elsewhere in the current sequence or any other sequence.

Navigating the Timeline

Before we leave the rough edit phase, let's take a look at a few different commands techniques you can use to quickly navigate around the Timeline.

Zooming In and Out

Though the zoom bar can be a quick way to zoom in and out, a quicker way is to do so via the keyboard. Avid has three keyboard commands that allow you to zoom in (Ctrl/Command+]), zoom out (Ctrl/Command+[), and see the entire sequence (Ctrl/Command+/). These are all designed to be accessed via the right-hand's ring or pinkie finger. They work well, but as we often have our right hand on the mouse, their placement is awkward. And if you're a left-handed mouser, their position is actually rather difficult to get to using only the right hand.

These commands live in the Timeline fast menu and I recommend remapping them to a more convenient place on the keyboard. For the left hand I map them to Shift+Z, Shift+X, and Shift+C (Show Entire Sequence, Less Detail, and More Detail, respectively), while for the right hand I map them to Shift+<, Shift+>, and Shift+/ (Less Detail, More Detail, and Show Entire Sequence, respectively). This places them on the lowest row of the keyboard, just next to the Shift key. (I've yet to find a more useful position for them on the keyboard.) I've also experimented with using multibutton mice for these commands and currently use the two side buttons found on Logitech and Microsoft mice for zooming in and out.

Jumping In

I refer to this set of commands as "jumping" rather than zooming, as they are typically used to quickly zoom in on a specific section of the timeline.

- *Focus:* The Focus button (the H key) is especially useful for troubleshooting small problems like flash frames because it is a one-step zoom to a preset amount to analyze a small section. This command is a toggle, so pressing it again takes you back to where you were.

- *Jump In:* Ctrl/Command+M provides you with a unique-shaped cursor that lets you quickly jump into a specific duration in the timeline. Simply issue the command and lasso across the section you want to see and the system zooms in so that section fills the entire timeline. This technique is very useful for pinpointing a segment that needs more refinement.

- *Jump Back:* Ctrl/Command+J sends you back to the exact zoom level you were at before the jump in. These two commands allow you to quickly hop in and out, and I find them extremely useful for popping into check for sync, offline material, etc. Both of these commands also live in the Timeline fast menu and can be remapped, if desired.

Unwrapping Wrap Around

If you used Xpress Pro on a laptop or a large monitor you were probably driven partially insane by a devious option called "Wrap Around." Whenever you zoomed in, this option would use the available vertical room in the timeline to display your tracks as if they were staves of music on a sheet. Though this seemed like a great idea to the person who invented it many years ago, in reality this option has perhaps befuddled more users than any other feature in the system—especially because Xpress Pro users inexplicably couldn't disable it!

Fortunately this can be disabled in Media Composer via the Timeline fast menu. Simply uncheck the "Wrap Around" option. This feature is now disabled by default when you create a new user in version 3.0.

2

ZEN AND THE ART OF TRIM

"Every block of stone has a statue inside it and it is the task of the sculptor to discover it."

—Michelangelo

A sculptor has many tools at his or her disposal. Regardless of the medium used there are always those tools used to rough out the shape that remove large sections of unneeded material and finer tools used to refine and give life to small details. I like to think of Trim as the fine sculpting tools. Though you could certainly complete a sculpture using only the large rough removal tools, most mediums require the fine tools to give the sculpture the definition it needs to be truly considered a work of art. Not only that, but using only the rough-out tools makes any fine detail work extremely difficult and inefficient. The same holds for editing. Chapter 1 described and defined the rough to medium tools for sculpting a story. Trim is your set—yes, set—of fine work tools.

I've named this chapter "Zen and the Art of Trim" because I firmly believe that Trim is the heart and soul of the Avid editing system and what ultimately continues to set it apart from other systems. I've heard many editors over the years proclaim that "nothing trims like an Avid." They say so not just because of the trimming tools available to you, the editor, but because of the *way* Trim works in Avid. A good friend of mine after learning the "deep" trim approaches in the system exclaimed that it really felt like he was "one with his footage."

I certainly don't promise you'll experience such a revelation, but hopefully by the end of this chapter you'll have a better understanding of why so many editors feel the way they do about the system. Master Trim and you have mastered the system and changed the way you think about editing forever. Trim is creative, not just corrective. And I will promise you one thing: If you take the time to practice and integrate the techniques in this chapter into your daily editorial work, you *will* become a faster and more efficient editor.

Thinking Nonlinearly

Beginners bring linear thinking to the trimming process. This is potentially the biggest mistake you can make, short of deleting all your media. The first place I see this is when beginners misuse the Match Frame button. Think of this really as the "fetch" button because the Match Frame name is too close to the function that linear tape editors have been using since the beginning of computer-controlled timecode editing.

The traditional tape method is to get the edit controller to find the same frame on the source material as where you are parked on the master tape. The source tape cues up, you adjust the video levels to match what is already on the master tape, and then you lay in a little more of the shot, usually a dissolve or another effect. You can do this in Avid as well. This is logical and simple, but it completely misses the point. Every master clip that you add into the sequence is linked to the rest of the captured material. You don't need to go get it because it is already there. Think of the extra captured material as always being attached to every edit in the timeline all the time. Each shot in a sequence is a window onto the original source material. The window can be moved, enlarged, contracted, or eliminated in the sequence, but the original source material is still there. It should be used for reviewing material, not as an integral part of the trimming process. However, if used the incorrect way, it is another dog paddle.

The best way I have discovered to think about trimming is to imagine moving earlier in time or later in time to see a different part of the shot. Coincidentally, as you move earlier you may be making a shot longer or shorter. Any trim that adds or subtracts frames—any trim that is on one side or the other of the transition—changes the length of that track and must have a corresponding change on all of the other tracks in the sequence. Not all trims change the actual length of a sequence, but the ones that do—the trims on one side of the transition or the other—knock you out of sync if you don't pay attention. This means you must look to the tracks that are highlighted when you decide to add a little video. Don't make the beginner's mistake of thinking that just because you are adding a few more frames to lengthen an action, it is a video-only trim. All the soundtracks must be trimmed if you make the sequence longer or shorter in any way.

The main reason that trimming is so much better than just extracting the shot and splicing it back in is that you have the immediate feedback of seeing the shot in context. When you use Trim to fix a shot while it is in place you get that instant sensory feedback that is so important when using a nonlinear editing system. When expanding your use of the Trim mode, stay in sync as much as possible. Now obviously, there are times when you want to go out of sync, for cheating action or artistic purposes—I'm not talking about that. I'm referring to the skill of understanding the relationship between what tracks are highlighted and what kind of a trim you are doing. Some people get so flustered the first few times they try trimming with sync sound that they abandon it altogether and invent elaborate workarounds that are easier for them to understand. Lots of energy, not much style. This is one of those skills that film editors (those who have actually touched celluloid) have over video editors. It is pretty hard to knock yourself out of sync with a tape-based project, so thinking in terms of maintaining sync is quite foreign. But film editors must learn that whenever they add something to the picture—a trim or a reaction shot—they must add a corresponding number of frames to the soundtrack.

Staying in Sync

The easiest way out of this dilemma is to turn on the sync locks. The sync locks allow the system to resolve certain situations where you tell it to do two different things: make video longer and don't affect the soundtracks. The system adds the equivalent of blank mag (silence) to the soundtrack. This may be safer than trimming and accidentally adding the director shouting "Cut!" but it will also leave a hole that must be filled in later. Blank spaces in the soundtrack are really not allowed! You will find yourself having to return and add room tone or presence so the sound does not drop out completely.

There will be a time when you tell the system conflicting things. You tell it to make the video shorter, don't change the audio tracks, and stay in sync. This is beyond the laws of physics. In this case, the system cannot make the decision for you where to cut sound in order to stay in sync, so it will give you an error beep and do nothing.

Sync locks work best if the majority of your work is straight assembly with little complex trimming. It is very effective, however, when you are sync locking a sound-effect track to a video track. The crash and the flying brick need to stay together. Also sync locking multiple video tracks together to keep them from being trimmed separately may keep you from unrendering an effect.

Sometimes you will be cutting video to a premade soundtrack. The video and audio parts of the sequence do not give you sync breaks when you change their relationship. Here, you must be even more conscious of maintaining sync. Don't fall into the trap of thinking that you can knock yourself out of sync now and later; when you get a chance, go back and fix it. Believe me, by the time you get the chance to go back, you will have created a situation that takes much longer to fix than if you did it right in the first place.

The most important aspect of trimming is to be aware of when you are going to change the length of the sequence by trimming on one side or the other of a transition and which tracks will be affected. When you grasp these points and overcome the fear of going out of sync, you will have a much more powerful tool and feel much more comfortable with the workflow concept of refine, refine, refine.

Sync Break Indicators

With the extra power of the nonlinear world (and film was the first nonlinear editing system!), there is the responsibility of keeping track of sync. The Avid editing systems do a pretty good job of telling you if the video and audio you captured together or auto-synced together (matching sound and vision from separate sources after digitizing) have lost their exact relationship. They are the white numbers called *sync breaks*. I think of them as a silent-white alarm that, when I see them ripple across my time-line, tells me I most probably have made a mistake. The only time I want to see sync breaks is when I have cheated action or I am dropping in room tone. Sync break indicators can be turned on and off via the Timeline Fast menu.

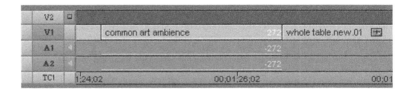

Trimming Fundamentals

Let's take a moment to look at some of the basic mechanics of Trim. I'm sure that much of this will be review for many of you, but you may also discover something that you didn't know or perhaps once knew but forgot.

Entering Trim

There are two basic ways to enter Trim mode:

- Park the position indicator near the edit point where you wish to trim and press the Trim mode button.
- Drag a lasso around the edit point where you wish to trim. When you drag a lasso, be sure to only drag it around the edit and not around any complete shots. Doing so will ensure that you enter Trim mode and not Segment or Slip mode (depending on whether you lasso from left to right to select a segment or right to left to slip a segment).

There's a third method worth mentioning, that has a very special use. When doing very complex trims you'll often take a few moments to select and enable the appropriate tracks and edits. If you hold down the Alt/Option key and press the Trim mode button, you can reenter Trim mode with all of the tracks and edits you had previously selected for trimming reselected. If you're doing complicated trimming—which you will at some point, I guarantee you—this technique is a huge time-saver.

Exiting Trim

Unlike entering Trim mode, there are many ways to exit, including switching to a different mode; but there are two specific ways that bear mentioning here:

- Click on the Timecode track. This is the most typical method used by the majority of editors I know. It has the benefit of not just exiting Trim mode, but also allows you to place the position indicator wherever you desire in the timeline.
- Use the Edit Review button. When you've finished a trim, you usually need to see the adjustments you've just made in context. The Edit Review button allows you to do just that with a single key. When pressed, the system exits Trim, moves backwards one edit plus two seconds, then plays. Think of this button as a predefined macro, similar to the Top and Tail functions mentioned in Chapter 1. This button is not mapped by default, but is available for mapping from the Play tab of the Command Palette.

Adding and Removing Trim Rollers

There are a couple of different scenarios where you may need to add or remove trim rollers. In the first, you wish to add or change the rollers on a track you've already selected, while in the second, you wish to add rollers to additional tracks in the sequence.

Though there are some common techniques, let's treat each one separately.

Adding and Removing from a Selected Track

Certainly the most common method used on a single track is to switch between A-side, B-side, and both sides trimming. This is easily accomplished using one of three different methods:

- Click on the A-side or B-side monitor to switch to that side or click between the two monitors to select both sides.
- Use the Trim Side buttons to switch sides. These keys are mapped to the P (trim A-side), [(trim both sides), and] (trim B-side) buttons, respectively.
- Use the Cycle Trim Sides button to switch between the trim sides. This button has the advantage of performing the function of all three of the Trim Side buttons but takes up only one key on your keyboard. Each key press toggles, in a loop, between A-side only, both sides, B-side only, both sides, and so on. This button isn't mapped by default, but can be mapped from the Trim tab of the Command Palette.

The Cycle Trim Sides button has the added capability of switching the green Audio monitor bar from one side to the other. If you have both sides selected for trim and want to switch the position of the green Audio monitor trim bar from one side to the other, simply press Cycle Trim Sides twice.

Another typical scenario to add trim rollers is to add similarly positioned rollers on additional tracks. This is most easily accomplished by simply enabling the desired tracks; the system will enable the nearest trim rollers it finds to the currently selected edit. The trim rollers on the newly enabled tracks will have the same side selected as the current track(s). You can also remove rollers from tracks by simply disabling the desired track.

Finally, you may want to simply add or remove rollers at specific edits on specific tracks. In these instances, just hold down the Shift key on the keyboard and click to add the rollers. The Shift key adds a roller to either the A- or B-side. If you want to add two rollers simply Shift+click on both sides of the edit or press the Cycle Trim button once. Similarly, Shift+click on an active roller to remove it.

Adding Rollers Where No Edits Exist

Let's take a look at the following editorial scenario. The sequence contains multiple video and audio tracks but you only want to trim on a subset of those tracks. Perhaps the timeline looks similar to the following illustration. You need to shorten V1, A1, and A2, but maintain sync across all tracks.

Though you could certainly use sync locks to achieve the fix, there is another approach that is extremely useful. If you hold down the Alt/Option key while you press the Add Edit button, edit points will only be added to those tracks that do not contain clips. This gives you an edit point to trim on where none previously existed. And, if you are in Trim mode when you issue this command, these new edits will automatically be selected for trim, as shown in the following illustration.

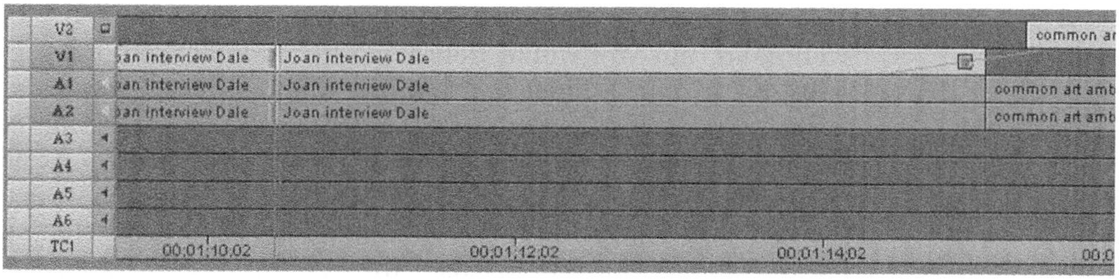

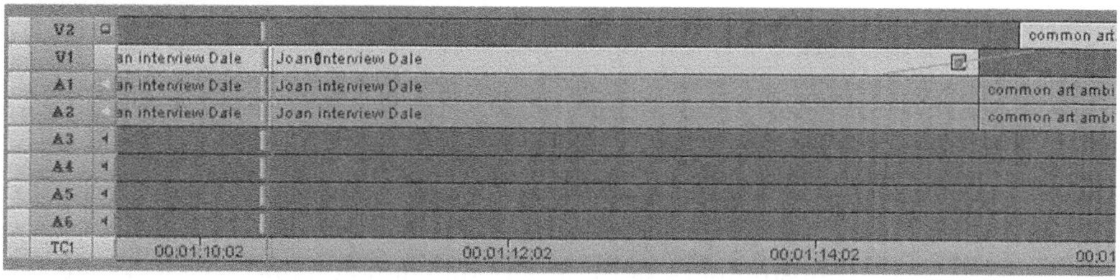

You can now trim in confidence that sync will be maintained across all tracks. At the end of your trim you can choose to either leave or remove these edit points. Personally, I like to remove them as soon as I've finished using them, but they can certainly be left in the timeline. Technically they won't cause a problem in your sequence but you can always choose "Clip > Remove Match Frame Edits" to remove them en masse in your timeline.

If you'd prefer to remove them at the end of the trim, simply press the Backspace key on the keyboard *before* you exit Trim mode. The Backspace key issued from Trim mode instructs the system to remove all selected match frame edits, selected edits being those that are currently being trimmed. You can think of this as a quick shortcut for the Remove Match Frame Edits command.

This function isn't a replacement for the sync locks but merely another method you can take to solve a complex sync problem. One limitation with this technique, though, is when you have a nonempty track without an edit at the same location, as shown in the following illustration.

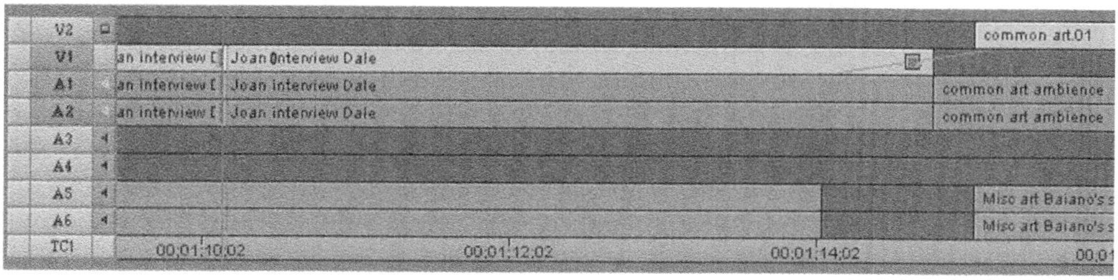

In this instance, you could still use the Alt/Option+Add Edit command, but you would then need to manually select a trim point on A5 and A6 before trimming, otherwise you would risk losing sync on those two tracks. This example is a great one for the benefits of sync locks. If you were to turn sync locks on for all tracks and then trim the tail of V1, A1, and A2, the gap between the two clips on A5 and A6 would automatically be tightened up for you and the timing between the cuts on A1/A2 and A5/A6 would be maintained, as shown in the following illustration.

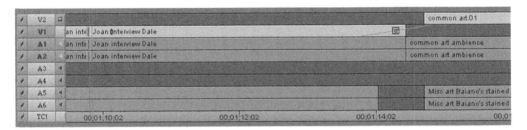

In summary, both Alt/Option+Add Edit and sync locks are key techniques to maintaining sync across multiple tracks in the sequence. Each has its distinct advantages and, arguably, disadvantages. (Sync locks really only show their power when you have a complex audio bed and/or lots of video effect composites. In simple timelines they can actually prevent you from making a reductive trim. This is probably where they have gained an undeserved negative impression with many editors. If you discarded them as a tool long ago perhaps it is time to revisit them!)

Methods of Trimming

Now that we've reviewed—and expanded upon—some of the fundamentals of Trim mode, let's look at all the different ways we can actually perform a trim. As with the previous section some of this will be familiar and hopefully some of it will be new. Every editor has their favorite method of trimming. Perhaps you'll discover a new favorite in this section!

Drag Trim

This is the most fundamental method of trimming. After selecting the appropriate trim tracks, edits, and sides, simply grab a roller and drag it to its new position. The video monitors in the Composer window will update as you drag, showing you the result of your trims. Be careful, though, to always drag *from* an existing roller, including the correct side! If you try to drag from an edit or side that does not contain a trim roller you'll remove all existing trim rollers and create a new one. Usually it only takes a few times of doing this to remember the rule. If you do a lot of drag trimming and don't want to be endlessly frustrated be sure to learn it and live by it.

Remember that the total number of frames you have trimmed on both the A-side and the B-side is indicated via the Trim numbers in the center of the Composer window, just above the command buttons.

One disadvantage to drag trimming is that you only see the picture change. If you want to hear the audio change you must hold down the Shift key (or use Caps Lock) to activate Digital Audio Scrub. Remember that this technique will play the frame of audio you are trimming as you drag the rollers. If you are trying to drag right up to the beginning of a sound bite you can use Digital Audio Scrub to hear the beginning of the bite—and perhaps even the breath before it. In addition, you can enable the audio waveforms by selecting "Timeline Fast menu > Audio Data > Sample Plot." As this is a relatively awkward menu command to get to, I strongly recommend mapping it to a key. If you don't already know how to do this, we'll discuss it in Chapter 3.

Power User: Defining Which Frame's Audio Is "Scrubbed"

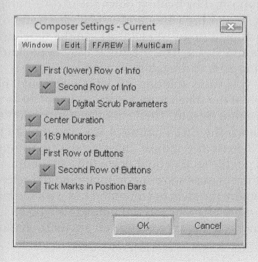

When you move the position indicator by one frame, there are two possible frames you might want to hear the audio from: the frame you park on or the frame that precedes it.

To see and modify the frame heard, you must display the Digital Audio Scrub parameters in the Composer window.

Once enabled, you will see either a 0/1 or a 1/0 in the outside corner above the Source and Record monitors.

- *0/1:* You hear the frame you are parked on.
- *1/0:* You hear the frame to the left of the frame you are parked on.

You will almost always use Digital Audio Scrub in the default (0/1) configuration, especially when you're scrubbing and marking. But when you are trimming on the tail of a shot, it may be preferable to hear the preceding frame instead. This is especially true if you are trying to remove a blip, breath, or similar sound at the end of a shot.

Let's take a look at the following scenario. The timeline below shows an undesired sound at the end of the clip.

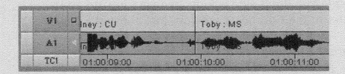

When you select the A-side of the edit and trim backwards you want to hear that the undesired sound is gone *when* it is really gone. If Digital Audio Scrub is set to "0/1" you will hear the sound from the frame you just trimmed away (and is therefore no longer in your sequence), but if it is instead set to "1/0" you will hear the frame that remains at the end of your edit. Give it a try and you'll see what I mean. I know editors who do tight audio editing all day who swear by this option and are switching it back and forth as they move through the editorial stages.

Note: If you want to use this in Trim there is a critical fact you must be aware of. The Digital Audio Scrub parameters above the Source *also* set the scrub configuration for the A-side in Trim. Likewise, the Record monitor *also* sets the scrub configuration for the B-side in trim.

Snapping to an Edit or Mark

When using drag trim you may often want to snap your trim position to another edit in the timeline or a mark. Both of these can be easily accomplished by holding down the Ctrl/Command key while dragging. When the command is held down the trim rollers will snap to all edit points on all tracks in the timeline. It will also snap to In and Out marks, making it easy to premark a position for trimming prior to dragging.

For example, if you were splitting an edit, you could play through that section of the timeline and mark an In or Out at the point where you wanted to split. Then simply enter trim on the video track, hold down the Ctrl/Command key, and drag to the mark you placed. Release the mouse and you will be trimmed exactly as you desired. (Be sure to hold down the Ctrl/Command key through the entire mouse movement, including the release, or the system will ignore the snap to command.) As we'll see later, there are arguably more efficient ways of doing this type of split edit, but this technique works very well if you are most comfortable with drag trims.

Keyboard Trim

Dragging is fine, but it tends to be a bit inaccurate, especially if you are trying to trim in or out a *beat* or trim a specific duration of time. For these instances, you might want to use the keyboard to trim instead. Two different types of trims are available: offset trims and directional trims.

Offset Keyboard Trims

These types of trims use the numeric keypad on your computer. Naturally, these types of trims are easy to do if you have a full keyboard but they are a bit harder to do on a laptop or other reduced keyboard, at least with one hand. To perform an offset trim you type the number of frames you wish to trim, indicate the direction you wish to trim using a + or – symbol (a "+" indicates a forwards trim while a "–" indicates a backwards trim), then press Enter. You can issue these in any order, as the commands "+15" and "15+" will both accomplish the same thing: a 15-frame forward trim. You can even change your mind and switch from forward to backward or vice versa by typing the other modifier before you press Enter.

Be sure to remember the rule from Chapter 1 regarding numeric keyboard entry: one or two digits equal frames while three digits equals seconds and frames. If you want to trim backwards by, for example, 120 frames, you need to type "–120 f" then press Enter to tell the system to count by frames, not by timecode (seconds and frames). Unlike the "+" and "–" modifiers, the "f" modifier must *always* follow the numbers.

Directional Keyboard Trims

These trims use the trim keys on the numeric keyboard. These keys will trim the selected edits one or ten frames backwards or forwards. You can also use Digital Audio Scrub in conjunction with the trim keys, just as you can with drag trimming, or turn on the audio waveforms. If you are using the single-frame trim keys this method can be very useful in trimming up to the beginning or end of a sound or breath.

Both the offset and directional techniques are great if you know—or feel—the amount of frames you need to trim, for it is far quicker to press the M key once to trim backwards by ten frames (or type "–10" from the numeric keypad) than it is to drag exactly ten frames. These techniques are great for opening up or tightening by "beats." Of the two options I often find myself using the numeric keyboard more than the trim keys, but that is possibly because I use the trim keys for on-the-fly trimming, as I'll describe in the next section.

Trimming with J-K-L Scrub

This is my second-favorite method of trimming on Avid. Just as J-K-L is a great way to find points in your source material, using it in Trim is a fantastic way of fine-tuning an edit. Indeed, many "old-school" editors like to think of J-K-L Scrub as "rocking reels" for the analog audio sound is analogous to the sound and precision one got by manually scrubbing on an open-reel audio tape deck.

J-K-L functions in Trim identically to the way it functions in Source/Record, with the exception that you are actively trimming material while playing. To review, Table 2.1 lists how to access the various play modes. (*Note:* If you have remapped these commands, press those keys instead.)

Table 2.1 J-K-L Trim Functionality

Operation	Key Usage
Trim forward at sound speed	Press L key
Trim backward at sound speed	Press J key
Pause playback	Press K key
Trim forward at faster than sound speed	Press L key twice for 2×, three times for 3×, four times for 5×, five times for 8×*
Trim backward at faster than sound speed	Press J key twice for 2×, three times for 3×, four times for 5×, five times for 8×*
Trim forward at quarter-speed	Hold K key, then press L key
Trim backward at quarter-speed	Hold K key, then press J key
Trim forward one frame	Hold K key, then tap and release the L key
Trim backward one frame	Hold K key, then tap and release J key

*The sound only plays at speeds up to 3×. Once you hit 5×, the sound, thankfully, cuts out.

Just as is the case with J-K-L play, the real power in J-K-L trim comes from the fact you can dynamically switch between all of the above trim. When using this technique it isn't unusual for an editor to "overtrim" slightly more than required then use J+K or K+L to roll back and forth until he or she has nailed the desired edit timing.

Trimming a Split: The Watch Point

If you wish to trim an already split edit, you can use J-K-L to trim either at the video edit or at the audio edit. Your decision will affect whether you are more interested in the picture or the sound transition. When you've selected a split for trimming, the blue position indicator will be positioned on the edit that will be used as the center point for your J-K-L trim. By default the position indicator is aligned with the video edit. I typically find that when I'm trimming a split I need to fix an audio problem, so I prefer to center my J-K-L trim at the audio edit.

Doing so is quite easy. Simply click on the trim roller you wish to center on and the position indicator will jump to that position. Unfortunately, there is no keyboard-based method of doing this—you must click with the mouse to move the position indicator. Moving the position indicator to the audio edit in a split also changes the center point around which trim loop play is performed.

V2	□		joan	
V1		16	Joan interview Dale	
A1	◄	Dale wreathes pictures 1C	Joan interview Dale	
A2	◄	Dale wreathes pictures 1C	Joan interview Dale	
TC1		00;01;00;02	00;01;02;02	

Trimming On-the-Fly

In my opinion, the best way to trim on an Avid editing system is to trim on-the-fly. Indeed, this is the very trimming technique that the editor I mentioned previously was referring to when he said that trimming made him feel one with the footage. When I trim I often start with J-K-L to get close then switch to on-the-fly to really nail the edit.

Trimming on-the-fly is a technique used during trim loop play. During the trim loop you can use the Mark In, Mark Out, or the keyboard Trim keys to trim the selected edit or edits. The Mark and Trim buttons operate differently while trimming on-the-fly, so we'll cover them separately.

Mark In and Mark Out

While the trim loop is playing, you can press Mark In or Mark Out to immediately update the edit to the point you marked. For example, if you are trimming the A-side of an edit and wish to cut out on that side immediately after a specific word in the dialog, simply press either Mark button at the desired time. The trim will be immediately applied and the play loop will begin again, looping through the newly trimmed edit. You can press either Mark button again to further refine the edit then see your changes in the next loop. And you can even open the edit up by pressing the mark button either past the edit (in the case of an A-side trim) or prior to the edit (in the case of a B-side trim).

As you can imagine, the interactivity is what makes this technique so fantastic and powerful. Instead of dragging, playing, stopping, dragging again, playing, stopping, and so on, you simply play, mark, immediately review your change, and continue to revise while playing as required. If you do the majority of your trimming by either dragging or entering numbers on the keyboard, then give this technique a try. You may well find your new favorite method of trimming!

If you are trimming on a single edit point the Mark In and Mark Out buttons have the exact same function. But if you are trimming a slip or a slide, the two buttons operate differently:

- Use the Mark In button to change the edit timing of the first, or left, edit.
- Use the Mark Out button to change the edit timing of the second, or right, edit.

Keyboard Trim Keys

As opposed to the Mark buttons, when you press the Trim keys on the keyboard the key presses are *not* immediately applied, but are instead accumulated and applied at the end of the trim loop. This enables you to press, for example, the comma three times to trim backwards three frames without interrupting the loop playback. When the loop completes, Avid will apply the three-frame trim and begin the loop again.

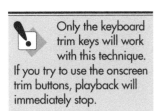
Only the keyboard trim keys will work with this technique. If you try to use the onscreen trim buttons, playback will immediately stop.

In my opinion, this technique is the finest "tool" in the Trim toolbox for it lets you trim away frame by frame and instantly see the result. When you're trying to fine-tune a dialog edit I believe there is no better tool to use. J-K-L would be a close second, but, as I mentioned earlier, I often begin the fine-tune process with J-K-L then switch to keyboard trim on-the-fly to really nail the edit.

Changing the Trim Loop Duration

By default, the Avid system uses a four-second trim loop and plays from two seconds before the edit being trimmed, known as *preroll*, to two seconds after the edit, known as *postroll*. (In the case of a slip or slide, the loop plays two seconds before the first edit, through the slip or slide, then two seconds past the second edit.) This can be modified two different ways.

The first way is to use the preroll and postroll fields in the left side of the command region of the Composer window. These fields are displayed if you have both rows of buttons displayed. This is easily accomplished using the Composer setting. (We'll discuss configuring the Composer window in Chapter 3.) Simply enter the duration desired for the preroll and postroll and you're set.

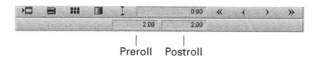

Preroll   Postroll

Another method—and one that can be accessed completely via the keyboard—is available via the Trim settings. The first tab in this setting provides not only the preroll and postroll settings, but also an intermission setting that can be used, if desired, to pause the loop. One possible use for the intermission is to give the client a chance to digest the loop before beginning it again. Personally, I don't use this setting, but I do know editors who do.

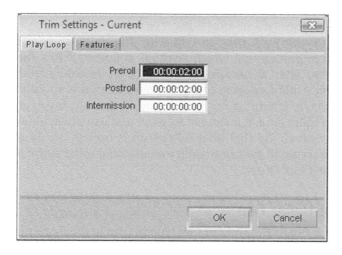

So how do you access and modify the Trim setting completely from the keyboard? Simple! Just press Ctrl/Command+4 to select the Composer window, then press Ctrl/Command+= to open the Composer window settings. Once the dialog is open you can use the Tab key to move through the three fields to enter the desired duration (in seconds and frames). Finally, just hit the Enter key to close the settings dialog.

As this is a user setting, your preferred trim loop preroll and postroll settings will be stored with your user setting.

Switching Trim Types On-the-Fly

While in trim loop play you can use the Trim sides or Cycle Trim keys on the keyboard to toggle the trim between an A-side, B-side, or both sides trim. After pressing the key, the trim will update and the loop will immediately restart. You can also use the keyboard track buttons to enable or disable tracks, but using these buttons will stop trim loop play.

Types of Trim

Now let's take a look at the various types of trim. We'll start with a quick review of single- and dual-roller trim then move onto the more advanced types of trim.

Dual-Roller Trim

Dual-roller, or center, trim is usually the first type of trim an editor discovers because it is has the lowest risk of knocking a sequence out of sync. But it is also the least-useful type of trim as it always does two in-kind edits. I've watched beginning editors roll back and forth over an edit with dual-roller trim as if they were trying to decide which "wrong" sound edit was the least offensive. That is because the chance of revealing undesired material while removing other desired material is so easy to do when you're trimming both sides simultaneously. If you're stuck in dual-roller land, it is time to step out and use single roller.

This isn't to say that dual roller trim is useless. Far from it! If you are trying to split or unsplit an edit, dual roller is the perfect trimming tool to use.

Single-Roller Trim

This is the most fundamental type of trim in the Avid system. It is also the type of trim that often scares beginning editors away from trim. Unlike other editing systems, the Avid system is "sync unlocked" by default. If you want to perform a single-sided trim on just the video or just the audio of a sync sound clip, the system will let you—even though doing so will knock you out of sync. Remember that, at the very minimum, you can press Undo to get out of any situation.

Slip

Beyond the basic situations, most of the difficult sync problems are fixed by using the Slip mode. First you have to ignore the "fact" that trimming a shot must make the sequence longer. Trim in the center of the transition, basically not affecting the sync, then use the Slip function. Many people have a difficult time grasping slip trim because it is so tied to the nonlinear, random access concept.

You can enter the Slip mode by multiple methods. With Media Composer and Symphony you can double-click on a clip once you are already in the Trim mode (as long as the timeline view allows you to see a black arrow cursor). In all models you can also get to slip trim by lassoing the entire clip from right to left. You may need to hold down the Alt/Option key to select the exact clip in a complex timeline. I generally get there by double-clicking in Trim mode because I use it as a second step in a difficult trim situation.

Think of slipping as a shot on a treadmill. The shot slips forward or backward, showing an earlier or later part, but the place in the timeline never changes. A slip will change the content of a shot by revealing new material, but leaves the duration of the shot and location in the timeline the same. Because you usually have more video linked to any shot used in the sequence, you can slip that entire shot back and forth.

So if you trim the beginning of the shot ten frames as part of a center trim, you can slip the shot back into position so that it still starts with the same frame. If the first ten frames of shot B are important, then slip them back into place. Your center trim moves the frames viewed in the A and B shots of a transition to be ten frames later. Although shot A gets longer and shot B gets shorter, the length of the sequence is not affected and the sync is not disturbed. Shot B has gotten shorter, but after you slip, it still has the same starting frames. I have worked with producers who have edited their programs on Avid systems for years who had never seen slip trim! Although it seems complex at first, it is truly a powerful tool when used in the right place.

Another very powerful way to use Slip mode is to use it to search through B-roll footage. Let's say that a piece of B-roll you've edited into your sequence just isn't working for you. You know you want to use B-roll at that point in the timeline, but the shot just isn't working. Before you go digging through your bins looking for another shot to use, select the shot and enter Slip on that shot. Then press the L key (or J key) multiple times and start whipping through the B-roll clip. You may find another section of the same B-roll clip that works better than what you cut in. And the beauty of searching through your clip via Slip mode is that once you've found the footage you're happy with you're basically done. A little fine-tuning to get the right first frame is all that is required.

Slide

The corollary to Slip mode is Slide mode. You can enter Slide mode by Ctrl/Option dragging from right to left, or Ctrl/Option double-clicking in Trim mode.

Slide mode moves the shot neatly through the sequence by trimming the shots on both sides of the selection. It affects the location of the shot in the timeline, but not the content or the duration. It is a good alternative to dragging a shot with the segment mode arrows if you are making a smaller change, but not nearly as useful as Slip mode.

Trimming in Two Directions

Trimming in two directions, an asymmetrical trim, is a subtle and very powerful technique. I typically use it when joining two scenes together as it allows you to easily tighten up the transition between the two scenes while simultaneously carrying over audio from the first scene into the second. It also is useful when you must trim a video clip longer, but don't want to add extra material to a sound effect or music that would keep you in sync but ruin an edit. To help you grasp this technique, let's look at an editing scenario.

You are joining two scenes, B and C, together. To marry them so they appear to be part of a continuous story you want to overlap the audio from scene B with the video and audio from scene C. You don't want, however, to extend the audio from scene B as the director yells "Cut!" almost immediately after the last frame you used. To prepare for the audio overlap, you move scene B's audio to a separate track, as shown in the following illustration.

Now let's analyze what we want to accomplish via Trim. We want to shorten the tail of shot B/1-A's video but not affect the video (or audio) of shot C/3-X. This suggests a single-roller A-sided trim on the tail of B/1-A. But we don't want to remove any frames from the audio of B/1-A. We have to perform some sort of reductive trim, though, on A2 or we'll break sync downstream.

An asymmetrical trim means that we can trim on the head of one edit and the tail of another. The two trims will both either trim out or trim in footage, but roll in separate directions to accomplish this. Since we don't want to affect shot B/1-A's audio, we can select a B-side trim on the filler just past the audio clip.

Finally, we need to consider A1. We know we don't want to trim away any of C/3-X's audio but we need to perform a reductive trim on this clip as well or we'll lose sync. If we had sync locks on, this trim would be handled for us automatically. But if we want to do it manually all we need to do is select an A-side trim on the filler preceding shot C/3-X's audio on A1. These trim rollers are shown in the following illustration.

V1	B/1-B	B/1-A	C/3-X
A1			C/3-X
A2		B/1-B	
TC1	2:00	01:00 54:00	01:00 56:00	01:00 58:00

Now that the trim rollers are properly configured we're ready to trim, right? Well, if we are going to use drag to trim, the answer is "yes." But if we want to use any other trimming technique the answer is "not necessarily." Why? Simply because we will have trim rollers moving in two different directions. Take, for example, the concept of a J-K-L trim. If we were to trim backward using J+K on V1 and A1 the shot would be shortened. But if we were to trim backward using J+K on A2 the filler would be lengthened. It is critical that we tell the Avid system which roller we want to "control" so the trim operates as we expect it to. To tell Avid which roller you wish to control simply click on the desired edit. Clicking on either V1 or A1 means that a J+K trim will be a reductive trim. Clicking on A2 means that a J+K trim will extend the edit.

Regardless of the roller you select, when you actually perform the trim you'll discover that the system will move all rollers in the correct direction so that the overlap is created and everything stays in sync.

V1	B/1-B	B/1-A	C/3-X
A1			C/3-X
A2		B/1-B	
TC1	2:00	01:00 54:00	01:00 56:00	01:00 58:00

Trim Two Tails (or Two Heads)

Version 3.0 introduces a powerful new method of trimming: trim two tails or two heads. To help you grasp this technique, let's look at an editing scenario.

You are well into the edit and need to shorten clip B/2-X in the sequence illustrated below. Unfortunately, the duration of the scene is already locked. What are your options? Well, you could use a dual-roller trim at the edit point between B/2-X and B/1-A, but that would change the head frame edit on B/1-A, which is undesirable. Alternatively, you could slide B/1-A, but that would change the head frame edit on BA/3-X, which is also undesirable.

Since neither single trim approach really works for you, you'll likely do an A-side trim on the tail of B/2-X, write the number of frames trimmed on a piece of paper, then find the tail of another shot in the scene that you could extend. It works, but hopefully you won't get distracted by a panicked producer while you're searching for that other clip.

In Media Composer 3.0 we've provided a new trimming technique that not only solves this problem, but many other similar problems. Indeed, I may never do another slide trim again. You can now select two A-sides (tails) or two B-sides (heads) anywhere in the timeline and perform an asynchronous trim on those two edits! In this scenario let's select the tail of B/2-X and BA/3-X, as seen in the timeline below.

Note the trim rollers. To make the above selection I lassoed the edit between B/2-X and B/1-A, switched to an A-side trim, and Shift + clicked on the other two rollers. Once selected, you can use any trim technique you desire (drag, J-K-L, on-the-fly, etc.). After trimming the tail of B/2-X, the timeline looks like the following.

Notice that the position of shot B/1-B (and everything afterward) has not changed. You made your adjustment in one interactive trim without changing the duration of the scene. As you can see, this is a very powerful trim technique and one that will likely change the way you approach some complex trimming situations, especially late in the game, editorially.

Trimming in Filler

There will be times when you'll enounter a trimming situation when you want to remove material from only one side of an edit but you want to maintain the duration of the section of the sequence. For example, you may have a synced clip with some undesired audio—such as an audio pop or an off camera thunk—at the head or tail of the edit. You're happy with the duration of the cut, but need to get rid of the audio. If there's no obvious sync in the clip, then you can often solve this problem by slipping just the audio—but sometimes slipping is just not possible.

In these instances you can hold the Alt key down (Windows) or the Ctrl key down (Macintosh) while performing a single-sided trim. Instead of simply reducing the duration of the selected clip, frames of filler are edited into the sequence for every frame of footage removed from the clip. You can use any trimming technique desired with this technique, but be sure to hold the modifier down until you complete the trim—especially when using J-K-L trim.

If you use this technique with audio you may find yourself with a location in the sequence where there's no audio playing. This absence of sound will be very obvious to the listener and you should replace the filler with recorded silence from the environment of the shoot (often referred to as "room tone") so that any atmospheric noise present when the sound was recorded carries over to this region of your soundtrack.

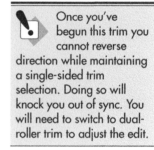

Once you've begun this trim you cannot reverse direction while maintaining a single-sided trim selection. Doing so will knock you out of sync. You will need to switch to dual-roller trim to adjust the edit.

Multicamera Take Names

In the previous examples the clips used a common naming convention for multicamera live sitcoms or dramas. The clip names break down as: Scene/Take-Camera. Scenes are numbered alphanumerically starting with A. If there is a camera reset and a scene continuation, the scene number is usually given a second digit (e.g., scene B and scene BA). In a four-camera stage shoot the cameras are typically labeled A, B, C, and X (for eXtra). Depending on the director the cameras can be positioned audience left to right in A, B, C, X or X, A, B, C order.

Trimming Outside of Trim

Let's conclude our discussion of trim by looking at two techniques you can use to trim outside of Trim mode. Some folks like to refer to these two methods as "trim unplugged," which is certainly an apt name.

Extend

As I mentioned earlier, the most typical use of dual-roller, or center, trim is to perform a split edit. As split editing is a very common technique, Avid includes the ability to do this outside of Trim using a function called Extend. Extend uses either a Mark In or Mark Out to indicate the direction for the Extend. The key to remember is that the edit you wish to extend must be contained *within* the mark. Therefore, if you wish to extend an edit backwards, mark an In point prior to the edit, and if you wish to extend it forwards, mark an Out point after the edit. Then simply turn on the tracks you want to extend, and press the Extend button. It will not knock you out of sync because it is a center trim and it trims both sides of the transition simultaneously. I find it most useful for mechanical trims that go to a direct and easy-to-mark point. It is best used, for instance, to extend a B-roll shot to the end of a sound bite. If any finessing is needed, I go to the Trim mode.

The Extend button is not mapped by default, but is available via the Trim tab of the Command Palette.

Slip in Source/Record

As slipping is also a commonly used function, especially to align disparate video and audio in sync, this function is also available outside of Trim. If you wish to slip a shot outside of Trim, simply park on the shot you wish to slip, turn on the tracks you wish to slip (and turn off those you don't), and use the keyboard Slip keys to slip the shot either backwards or forwards.

It is important to note that the trim keys in this type of slip function opposite to the way they do in Trim mode. Table 2.2 lists their functionality in Source/Record slip.

Table 2.2 Trim Key Slip Functionality

Key	Function
Trim left one frame	Slip forward in time one frame
Trim left ten frames	Slip forward in time ten frames
Trim right one frame	Slip backward in time one frame
Trim right ten frames	Slip backward in time ten frames

Notice that I use the term "in time" to describe the direction of the slip. If you were to park on a locator and slip the track containing the locator you would see the locator move left when you use the Trim Left keys and move right when you use the Trim Right keys. But in order for that locator to move left, the shot must slip *forward* in time. Think about it. For a locator to move to the left, the shot must start later in the original source. Starting later in the source means that you are trimming forward in time. I'll admit that it can sound confusing, but the best way to understand it is to try it yourself.

A typical way this is used is to park on an audio sound cue then slip the video until it aligns with the cue. I use this technique quite often when adjusting the timing of inserts. If the insert must be synchronized to a sound on the main audio track, such as, for example, the clinking of two glasses, I can drop the insert in over the sound, park on the sound, then slip left and right until the action in the insert is synchronized.

I've also used this to realign individual shots that drop out of sync and even to force audio out of sync to remove an undesired sound. Remember that you can always break sync if the sync break won't be noticeable to the viewer. If there isn't lip flap or other obvious examples of sync you can slip the audio and video out of sync to remove an undesirable sound, such as an off-camera tap or thunk.

Conclusion

As you can see, Trim is an extremely powerful set of tools that you can use to refine and fine-tune your edit. If you aren't using Trim yet, by all means jump in head first! And if you've only just begun to trim, use this chapter as the impetus to dive into the deep end of the trimming pool.

3

INTERMEDIATE TECHNIQUES

> *"'Tis skill, not strength, that governs a ship."*
>
> —Thomas Fuller

It may come as a surprise that most beginners of the Avid editing system make the same basic mistakes. I don't mean mistakes caused by the software being too difficult, but mistakes from working hard to grasp some fundamental ideas. They are sometimes crucial mistakes, like not knowing exactly what to back up and then trying to restore a project with no bins. Sometimes it is a subtler mistake, like not using the power of a new tool because "That's not the way I work." You may be missing a huge opportunity to improve your speed and understanding.

I have heard people say that Avid is difficult to learn and that the interface has a steep learning curve. This is only partially correct. You can be mousing around the screen in only a few hours and really editing by the end of your first day. But as with any professional tool, you want it to go faster and do more and the Avid interface rewards this, yet very few people use everything the software has to offer. If you have a particular task to perform, there are the tools designed to facilitate that task in a straightforward way. If you need something a little different, there is a lot of room for variations. The variations take the most time to learn, but are the most rewarding.

The most basic mistakes are made right at the beginning, when editors are still trying to learn how to navigate through their material. There is a lot of translation going on between where they want to go, how they used to do it, and the two or three techniques they know how to use. They end up settling for the dog paddle before they have mastered the breaststroke, the crawl, the backstroke, and the sidestroke. They will always poke along unless they unlearn the method that wastes energy and, let's face it, gets them there without much style.

Multiple Methods to Solve One Problem

Here is another fundamental tenet of using the Avid systems: You can use multiple methods, one after the other, to fix a problem. Editors who consider themselves novices may have acquired one or two methods that they use under all circumstances. The more experienced editor uses a method of trim to get close as fast as possible, depending on the particular timeline view and level of sequence complexity. The editor then reanalyzes the problem and easily switches to another method to put the final polish on the fine points. The "right" way is the way that accomplishes what you want in any given particular situation in the fastest way possible.

Using the Keyboard

You may have guessed by now that I am referring to the overuse and abuse of the mouse or trackball. This is where you should start to improve your technique. When you first learn, use the most obvious way—the mouse. This helps you get over the beginner's problem of trying to remember what you want to do next and where it is on the screen. After this beginner's stage, you instantly forget how magical it all is and want to go as fast as possible. Then you must cast down the mouse! Use your keyboard!

Force yourself to use the edit keys and keyboard equivalents as soon as possible. If you haven't put the colored keycaps or stickers on your keyboard yet, you are missing a whole world of speed. Look at the Ctrl/Command key equivalent for the functions you use the most and think up funny little ways to make them stick in your head. Ctrl/Command+Z to undo and Ctrl/Command+S to save should be comfortable before your first day is over. Then start to use Ctrl/Command+W to close windows and Ctrl/Command+A to select all. Use F4 to start capturing once you are in the Capture mode and, as an ongoing project, memorize the Tools menu. You should mark Ins and Outs mainly from the keyboard so you can keep your material rolling and mark on-the-fly. Trimming can be done in several ways from the keyboard. You don't want to give yourself a repetitive stress injury, though, so always try to make the environment as friendly as possible for your wrists. Get the keyboard at the right height, get a wrist pad if you need it, and give your wrists the rest time and exercise they need to keep functioning. Use the extra time gained with these keyboard techniques to watch the sequence one more time and think about it.

Customized Keyboard

There are many ways to personalize the keyboard. I don't recommend any special keyboard layouts since I believe they all should be created organically from observing the functions and keys you use the most. You can take any button from the Command Palette and put it on any key (button to button). Some recommend learning the colored keycaps and modifying only the function keys or the shifted functions of keys that make alphabetic sense to you. For example, you can use the shifted function of the keyboard to store a wide set of mnemonic-based shortcuts; put Render on Shift+R, Subclip on Shift+S, Import on Shift+I, Export on Shift+O, and so on!

Once you've mapped commands to your keyboard, don't be afraid to change them. If you find yourself not using a mapped command anymore, change it! There is no reason to leave a key mapped to a function you never use if you could better utilize it. One rule to consider is that if you find yourself consistently using a command for the menu in a session, map it! I know of some editors who use the function keys as their "session scratch pad" and remap them continuously to commands they are using at the moment. If the function becomes a consistent part of their daily use they then map it down to the main keyboard.

Over the years my keyboard has evolved with the type of editing I do. My current keyboard map is primarily customized on the left side. That is because I both edit on a curved "ergonomic keyboard" and I tend to leave my right hand primarily on the mouse. I also make extensive use of shifted commands as I find it a quick and convenient method to access the commands I need to use. The following figures show my current keyboard in both its normal and shifted function.

Some functions on the pulldown menus do not have keyboard equivalents or buttons. You will need to map them in order to use them as a single keystroke or in conjunction with a third-party macro creation program. A good example of this would be the Audio Data > Sample Plot (or audio waveform) command that lives in the Timeline Fast menu. If you find yourself toggling this on and off during your edit session it is far faster to have it mapped to a key than to navigate through the Timeline Fast menu every time you want to turn it on or off.

Here is how to map a pulldown menu to a key:

1. Open the Keyboard setting and the Command Palette at the same time.
2. Click on the Menu to Button reassignment button on the lower right of the Command Palette. Your cursor will change to an icon of a mini-pulldown menu.
3. On the Keyboard setting window, click the key you want to map. Hold down the Shift key if you want it to be a shifted function.
4. Choose the menu you would like to map from the pull-down menu choices. The initials representing that function will appear on the key.

When you are mapping the keyboard, be sure to save your settings. This is quickly accomplished by highlighting the Project window and pressing Ctrl/Command+S. You may want to locate your user settings (which are in an Avid Users folder inside an Avid program folder stored in either the shared files or documents folder of your system or, for older systems, within the Program Files folder), and save a backup copy of the settings on removable media.

Once you feel comfortable with where everything is on the screen, push yourself to resist the mouse and keep your hands on the keyboard! You may find that you need to create several key-board settings and use them for different types of projects.

If you have a custom keyboard, be aware that it works best for the version of the software you were using when you created it. If you go to an earlier version of the Avid software, your keyboard, like all user settings, may have features that do not exist in the earlier version. Usually you can go forward to the latest software version with Keyboard settings, but this has been known to occasionally create odd, unrepeatable problems. It is best to take a screenshot of your Keyboard settings (use a shareware screen capture program to save the Keyboard settings window as a graphic file), print it out, and make the custom keyboard again.

Navigating Nonlinearly

Another beginner's challenge is to not think linearly when jumping large amounts of time. Editors make a compromise and move at the speed of human comprehension in order to grasp a particular point in the material or listen to the performance. But what if you just want to get there as fast as possible? I see beginners actually dragging the timeline's blue bar through their material or sequence to get to the end! It's random access; get random. Jump to the end with the End key and jump to the beginning with the Home key.

Fast Forward and Rewind

The Fast Forward and Rewind buttons are useful if you just want to jump to the next or the previous edit, but these buttons are usually left to the user setting default of being "track sensitive." This means that the default for the Fast Forward and Rewind buttons jumps to the next edit that uses all the tracks that are highlighted in the timeline. If video track 1 is highlighted, then you jump, in a sequential way, from cut to cut on video track 1 only. Turn on all the tracks (Ctrl/Command+A with the timeline highlighted). Now you jump to every edit where all the tracks in the sequence have a cut in the same place. No straight cuts on all your tracks in the same place? With all your tracks highlighted, Fast Forward jumps all the way to the end of the sequence! That can be confusing since there is no easy way to get back to where you were in the timeline. There is no undo for jumping to the wrong place!

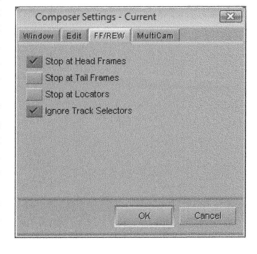

If you find yourself jumping to the end of the sequence a lot by accident, you can change the Composer user setting on Media Composer and Symphony so that it jumps to every edit regardless of which tracks are highlighted. You can change the settings to jump to every locator, too. You can reverse this default Composer setting instantly by holding down the Alt/Option key in combination with the Fast Forward or Rewind keys. This may be faster than going to the setting, especially if you need this option only occasionally.

Using the Timeline

Modifying Fast Forward and Rewind still misses the larger point, which is that you don't need to step through lots of edits just to get through the timeline. If you want to get near the end, just click there! But more important, you need to be able to see where you are going. Learn to change the scale or view of your timeline quickly and easily.

There is a marvelous and intuitive drag bar to resize the timeline. Drag the slider to the left and the timeline compresses; drag it to the right and it expands. You have fantastic fine control and can change the view by large amounts quickly with the same function. On Media Composer and Symphony there are more keyboard controls so don't neglect these:

- The Focus button (the H key) is especially useful for troubleshooting small problems like flash frames because it is a one-step zoom to a preset amount to analyze a small section. Since it is a toggle, pressing it again takes you back to where you were.
- The keyboard equivalents to the drag bar (Ctrl/ Command+[, Ctrl/Command+]), which I map on my keyboard to Shift+X and Shift+C, respectively, so they are always available under my fingers.
- Ctrl/Command+/ to show the entire sequence. This is always a quick reference if you get lost while zoomed in too far. I map this to Shift+Z on my keyboard.
- Ctrl/Command+J for "jump back."
- Ctrl/Command+M for zoom in or "more." Ctrl/ Command+M allows you to drag a unique-shaped cursor around a specific area in the timeline where you want to zoom in. This technique is very useful for pinpointing a segment that needs more refinement.

There are so many easy ways to change the scale of the view in the timeline because it is vital to using the power of the system. The timeline is not just a pretty picture. It is an important tool for navigation and should always be sized to fit your needs for that moment. Swoop in to do some fine trimming, step back a little and look at the whole section, then fly off somewhere else to fix the next problem. You should be considering the scale of the sequence view at every stage of the work.

Jumping Precisely

Even though you may see precisely where you are going, you may not always get there the fastest by just clicking the mouse. Don't dog paddle through the sequence; you need to combine the power of the random-access navigation of the timeline with the precision of the Fast Forward and Rewind keys. How can I jump huge distances in a single bound and still end up on the first frame of the cut? There are a series of modifier keys without which life as you know it could not exist. The most important is the Ctrl/Command key. When you hold down the Ctrl/Command key and click in the timeline, the blue position bar always snaps to the first frame of any edit on any track. It also snaps to marked In or Out points.

If you hold down the Ctrl/Command key while dragging your cursor through the timeline, it snaps to the head of every video and every audio edit. If you click anywhere in the timeline with the Ctrl/Command key held down, you are guaranteed to land at the head of the frame or a marked In or Out point. This can eliminate missing a frame here or there and creating flash frames. You must still combine this with the timeline zooming techniques, especially with a very complicated sequence. You may be surprised that you have snapped to the audio edit on track 7, which is three frames off from the video edit on track 1 you really wanted.

Lasso with Modifiers

This trick is one of the most important for using the timeline precisely and as a true nonlinear graphic tool. Holding down the Alt key on a Windows or the Control key on a Macintosh allows you to lasso any transition on any track and go into the Trim mode at a specific place. In combination with the Shift key, you can make multiple selections easily and, in a graphic way, extend the use of the timeline to get exactly what you want.

Changing the Track Name

A hidden timeline feature that is quite useful when working with lots of layers is the ability to name the track you are working on. You can right-click or Shift+Ctrl+click on the track number in the timeline. Choose "Rename track." Be careful, though. These customized names *do not* appear in the digital cut or audio mix tools. Instead these tools use the original track numbers.

Audio Monitoring

Many beginners don't see the connection between what they are viewing and why they can't hear the sound anymore. Being able to turn audio track monitors on and off selectively means you can concentrate on just the sound effects on track 5 or make five versions in separate languages and monitor one language at a time. But it also means you always need to keep an eye on which tracks are being monitored because they could be different from the tracks you are trying to use.

Occasionally, having all 16 audio channels playing at once is distracting when you just want to find the blip. Even though you may have 16-channel potential for monitoring, sometimes it is faster to solo a track for a critical trim. You can solo a track (or multiple tracks like a sound effect and a verbal cue) by Ctrl/Command-clicking on the desired audio monitor icons. This turns the entire monitor icon area green. To summarize:

- The audio track key will enable/disable the track for editing.
- Ctrl/Command-clicking on the audio monitor solos the track. Multiple solos are possible.

It is a significant workflow improvement to have all these tracks monitored when you are using J-K-L keys to fly through lots of material or a sequence where you have "checkerboarded" the dialog. Checkerboarding is a great dialog technique of putting each of two actors on separate audio tracks. This makes it easier to slip overlapping dialog without chopping off the previous spoken line. With one line on V1 and the next on V2, you can monitor something that sounds like the finished audio when you are trimming and continuing to tweak the sequence. No more lip reading at 2× speed!

Creating a Temp Mono Mix

During an edit session you may need to create a temp mono mix so you or the producer can hear audio clearly and are not distracted by an incomplete stereo mix. (It also eliminates the "Why is she talking out of only the left speaker?" questions from the producer.)

To create a temp mono mix:

1. Open the Audio Project setting and switch to the Output tab.
2. Click the Mix mode button until "Mono" appears. Now all audio clips will play in mono.

Keep in mind that this is a temp mono mix only. If you've already applied some pan and gain to clips in your sequence those adjustments will be maintained. Simply switch back to "Stereo" and your mix will be restored to its original configuration.

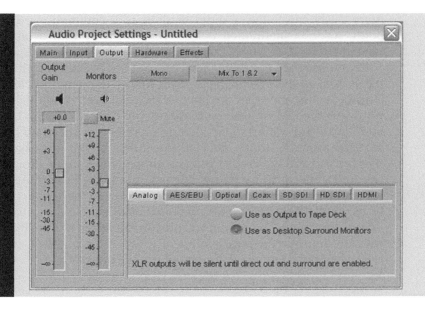

Organizing Your Material

One thing that working with a computer forces you to do is organize. If you don't have a plan from the beginning that is easy for you, you just won't do it, and you will find yourself relying on the frame view to find shots. Although this may make it easier for clients to put their fingers on your monitor and shout "That one!" it is slow. The bins that come from the telecine need to be named for scene and take, but after that, everything follows the standard script notation. In a traditional film edit room, organizing has more to do with tracking the physical film and keeping the right scenes ready for cutting. In a nonlinear edit, a film project must keep everything from a scene together. Documentaries or other formats where the form takes shape during the edit must live or die by good organizing of shots in a computer-based edit. I once edited a show with 200 sound bites and no script. To use the tools that are available for finding shots, you need to enter the information in a way that allows you to search for it easily.

There are two things to keep in mind when creating bins: tape name and bin size. Generally, the tape name is most important when you are starting to organize, because this is initially what you are handed from the field. Tape names should have some criteria for allowing you to trace back shots. For instance, with day and location coded into the number used for the tape name, you more easily can follow a series of shots to a common starting point. Also, as mentioned before, you can start with a tape-named bin for digitizing and then organize based on content.

If your bins are too large, however, you are defeating a lot of the benefit of the organizing. It takes longer to open and close large bins, and once they are open, it takes longer to find exactly what you need. Most likely, you will use the Find Bin button frequently, so you want it to be opening small bins to speed up the retrieval process. To use the Find Bin function, you must have that bin listed in the Project window (or in a folder in the Project window); it needs to have been opened once in the project. Also, you can go directly from the Sequence window to the source bin by using Alt/Option+Find Bin. Again, Match Frame is also a useful tool for calling up a shot if you are not interested in the bin. It calls up the source clip regardless of whether the bin even still exists! All Match Frame needs is media on the drives.

 If you mistyped the name of a custom column, simply hold down the Alt/Option key and click on the name to highlight it. Then simply type the corrected column name.

Using Custom Columns in a Bin

One feature that I'm continually surprised that experienced Avid editors don't know about is that you can create custom columns in a bin. Though there are dozens of statistical and metadata columns available by default, you can create your own custom columns by simply clicking on an open area in the Bin headings. Then simply type the custom name you want and start entering data!

Be sure to be consistent in not only how you name your custom columns but what you put in them. For example, if you create a custom column called "Shot Type" and enter both "Wshot" and "WS," you will have a harder time locating the shot you'll need. To facilitate accurate custom column entry, you can Alt/Option+click on a field in a custom column and a pop-up menu will appear containing all of the data you've entered into the custom column. Simply select the desired data and they will be inserted into the selected field.

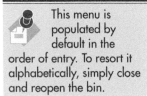 This menu is populated by default in the order of entry. To resort it alphabetically, simply close and reopen the bin.

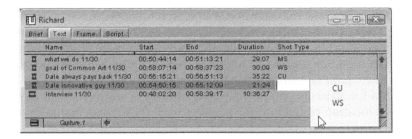

Sorting in the Bin

You can easily sort on any column by simply clicking on the Bin heading then pressing Ctrl/Command+E. If you wish to sort in reverse order (i.e., Z to A), simply press Ctrl+Alt/Command+Option+E. If you wish to sort by multiple columns, simply arrange them in the order you wish to sort, left to right, select all the desired columns to sort by, then press Ctrl/Command+E.

Sifting Clips

Though sorting often helps you find the clip you are looking for, sifting displays in the bin only those items that meet a certain criteria. Everything else is temporarily hidden from view. For example, you may want to display only those clips containing "INT" in the Location column. To access Sift choose "Bin > Custom Sift."

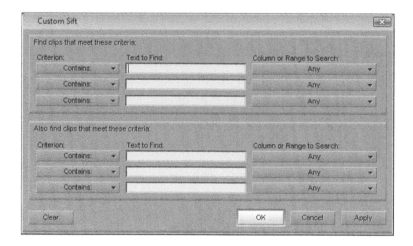

The Custom Sift dialog allows you to refine your search by up to three criteria. You could, for example, look for all Location: INT, Shot Type: CU, Scene: 47A shots. The second set of criteria allows you to search for a second set of clips, allowing you, for example, to also find all shots that are Location: EXT, Shot Type: WS, Scene: 47A.

You can even use this dialog to search by timecode. If the Start and/or End columns are displayed in your bin you can enter a specific timecode and find the shot that contains it. To do so, simply enter the complete timecode *without* the colons or semicolons into the find field then select "Start to End Range" from the search menu. I find this technique especially useful when the

To sift by film KeyKode number, enter only the numbers following the dash.

producer has screened footage using source or select reels with timecode burn-in. If he or she found a great line reading at 07:16:44:05, I can simply enter that timecode value in the appropriate bin—or even the Media Tool—and quickly find the clip containing the desired line reading.

There's one "gotcha" with the Sift function, though, that has burned many an editor. When you sift a bin you are hiding some of its contents. Don't make the mistake of opening a sifted bin and panicking because some of the shots are missing. Before throwing open the edit bay door and screaming out that when you get your hands on the so-and-so that deleted your media, pause briefly and look at the name of the bin. If you see the text "(sifted)" to the right of the bin name then the bin is in a sifted view and other clips, sequences, and so on could be stored in the bin. Simply choose "Bin > Show Unsifted" to show the entire contents of the bin.

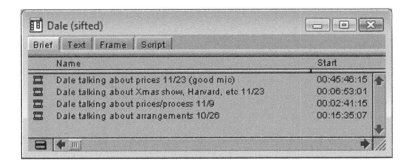

Using the Media Tool for Editing

The Media Tool is a good way to find shots, especially if the bin has been deleted, by sorting or sifting in the Media Tool window and then dragging the shots to a new bin. If the media files are online, you can get them through the Media Tool. Most people use it for media management, but it is an easy way to find hidden shots quickly by searching across projects.

On a very large job, or a system with lots of media, opening the Media Tool can be a time-consuming process causing some people to avoid it completely. That is why you can choose exactly what drives or what projects you will search through before you open the Media Tool. The smaller amount of media that are searched through, the faster the Media Tool will open. Opening the Media Tool causes the system to read the media databases from every active MediaFiles folder and load that information

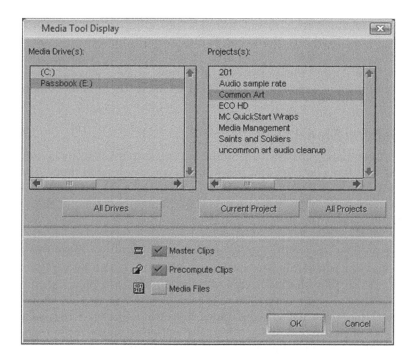

into RAM. That can take drive access time and potentially use up much of the available RAM (if you have hundreds of thousands of objects or several terabytes of storage). Of course, if you are really desperate to find a shot, you can just press All Drives and All Projects, and load the entire media database. Once loaded completely, all searches in the Media Tool will be instantaneous for the rest of the session or until you shut down the computer.

Early in the process of organizing a new project, if the Media Tool is not too large yet, I leave it open to avoid having to keep opening a bunch of separate bins when looking for a shot. Think of the Media Tool as the largest bin of media you can have at one time, and when it gets really large, open it only when you must.

So far we have discussed stylistic techniques; the benefit gained is primarily in speed, efficiency, and making you look good. But a few common techniques, if not followed regularly, can create real technical problems. I'll describe some of these problems in Chapter 12, but let's cover a few that can be prevented easily by just following directions.

Customizing Your Interface Environment

There has been so much progress in the user's ability to change and optimize the user interface that now we need to discuss the most effective trick to cut through the options. Most interface changes can be saved and recalled quickly through the judicious use of workspaces, toolsets, and custom views. The trick to finding and using your perfect setup is how quickly you can keep changing it to exactly what you need at exactly the right moment.

General Modality

A general philosophy for modal editing systems is to have only the functions you need in front of you when you need them. The idea of modes is very powerful and should not easily be dismissed by marketing hype. If you have all the functions available all the time, what is the possibility that you will need more than a very small percentage of them? The rest are wasted and clutter valuable screen real estate. If pressed accidentally, these unneeded functions may cause more harm than good, sending you into a function or display change that is not desired. Did you mean to trim that shot or just navigate there to look at it? A mouse click in just the wrong place will give the undesired result. A very careful mouse click in just the right place eventually will cause carpal tunnel syndrome. A nonmodal interface also may obscure the more needed functions at just the most critical time when you must have them close at hand. If you can't find a function, it doesn't exist.

A modal system gives you a series of streamlined, focused interfaces for the most used functions. To go to or from a mode should be seamless; this is where the real challenge comes. If you can't get to a mode easily, then you may feel that you need to have the important parts of that mode available all the time. The Avid editing interface is based on the Source/Record mode being a type of home page. By pressing the Escape key you can get to it instantly from any other mode. You can get to the other modes, trims, effects, and color-correction functions through dedicated buttons that can be mapped to the keyboard. These mapped keys are critical to using the modal interface to its most powerful advantage.

Custom Views

There are custom views for the timeline as well as for the entire user interface. You can change colors and track size, position, and information displayed for the timeline to display just the information you need; for instance, for audio mixing or effects creation. Create the view in the timeline, then click on the default name of the view and choose Save As. You can use the general user interface to eliminate or enlarge buttons and use color as a key to the functions you are using and sign as to what custom setting you are using. Go to Interface in the Project window, then the

Appearance tab. If you really mess up the view (and potentially can't see anything to change it back) go to that view and right-click/Ctrl+Shift+click and choose Restore to Default. When you are done customizing a setting, name it something useful so that you can tie it together with a workspace or toolset. If you have other user settings that are meant to be used at the same time, name the interface the same thing.

Workspaces and Toolsets

One way to conquer the complexity of modes and settings is to create a series of snapshots of your favorite configurations. These are workspaces and toolsets.

Workspaces can be created from scratch, whereas toolsets start with some preset modes like Source/Record, Color Correction, and Effects. You can modify both types to reflect your personal choices for button layout, screen colors, text, and button size. Think of these as user interface setup macros since they can even contain Project and user settings. If you find yourself switching between any two user settings on a regular basis, just program them into the workspace. If you need to have a timeline change the size, color, and information displayed when you start to do audio mixing, link the workspace or toolset to the particular timeline view.

To link a toolset or a workspace to a setting you must first create the setting and then give it a name. For workspaces the name of the linked setting must be the same as the workspace. For the toolset you can link a preset toolset to any user setting name. If you have multiple user or project settings that you want to change at the same time, you must give them all the same name. You can name workspaces and other settings by clicking the empty space to the right of the default setting name in the Project window. You can create multiple versions of any setting by clicking once to make it active and then using Ctrl/Command+D to duplicate it (the duplicate command is also under the Edit menu and is a right-click/Shift+Ctrl+click choice). Then change the setting and rename it. To see your new setting make sure that the Project window is displaying all settings, not just active settings. Otherwise you will continue to duplicate settings and never see them!

The following method links user settings to workspaces:

1. Create a timeline view that is designed especially for audio. Turn on important audio graphic information, make the video tracks smaller, and move the timecode track in between the video and audio tracks.
2. Save the timeline view and name it "Audio."
3. Open important audio tools that you like to use, like the Audio Mix and Automation Gain windows. Position them

where they are best integrated with the rest of the interface.

4. Click a workspace in the user settings window and duplicate it (Ctrl/Command+D).

5. Double-click to open the workspace.

6. Choose "Activate Settings Linked by Name" and "Manually Update This Workspace."

7. Click on "Save Workspace Now."

8. Click the empty space next in the user settings window on the workspace setting to name the workspace. Name it "Audio." If this is your only workspace then it is "Workspace 1." Workspaces are numbered based on alphabetical order in the user settings window. They can change numbers dynamically when you add new workspaces so be careful when you name them. You may want to name the workspaces with numbers like "1 Color-Correction Workspace" so that when you create an audio workspace the order won't change.

9. Open the Keyboard setting and the Command Palette at the same time.

10. Go to the More tab and grab the button for W1. This audio workspace is "Workspace 1." Map it somewhere you can remember easily like Shift-1.

When you press this button you will call up all the audio windows and the timeline will change. Make a workspace or a toolset in a similar manner for all your important functions. The toolset actually is easier since you can choose "Link Current to …" from the Toolset menu and get several options for linking to different user settings, although you have fewer toolsets to work with.

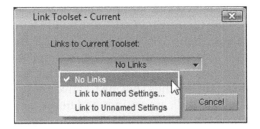

Backing Up

If you are a beginner to the computer, you may not realize the seriousness of backing up, but first imagine the cost of losing a day of work. Then imagine losing the entire project. Until you lose your first project, you might not back up on a regular basis. The fact that for about a dollar and five extra minutes you easily could have saved all the project information is a very compelling argument. There is nothing more sickening than returning to a project after a short break and not being able to find a sequence. Back up everything important, even twice a day, using rewritable CD-Roms. Always have your work somewhere else when something goes wrong. Auto-save can fail if the drive is too full and random crashes can destroy boot drives with projects and sequences. I back up constantly, but not to the same disk! Use a separate removable disk for every day of the week, and you will have seven chances to find a usable version.

What should you back up? It is usually difficult to back up all of your captured media in the middle of a project. But with terabyte and larger FireWire® and USB external drives becoming affordable, backing up media isn't out of the question. Backing up captured media may not be top priority in your project unless you are working with lots of nontimecoded material. Batch digitizing can be faster than any other form of restore—if you have the source tapes! What you really need to recreate your job is the project folder on the internal hard drive. Take all of it, not just the bins or the project icon. Many short-term projects should have a project folder that can fit onto a high-density floppy. If you are working on "History of the World" or need to back up lots of graphics and music tracks from CDs, however, you need to back-up to CD-Rom or DVD. A CD-Rom holds 650 megabytes of data and a DVD will hold 4.75 gigabytes.

Alternatively, you can send your project over a fast network to a system that contains a drive suitable for backup. On a particularly big job several years ago, I backed up to a USB drive on an Ethernet server every hour. Although we lost power almost every afternoon while editing on location, we never lost any data.

Backing up to another hard drive on your system or another computer at your facility may not be good enough. I have had both the Boston Fire Department and Mother Nature ruin two different suites where I was working. (To be fair, the fire department was trying to save an historic building.) A particularly successful film assistant I know in Los Angeles makes two backups of the project as the film gets close to picture lock. He takes one, his assistant takes one, and they take separate routes home. It is Los Angeles after all. Avid technical support has a category in their database to record reasons for equipment failure. Earthquake is one of them. And having a flash drive in your pocket also means you are more likely to get paid as a freelancer.

Nontimecoded Material

In the rush to complete a project, people throw anything and everything into their project just to get it done. Scratch audio is recorded straight to the timeline, VHS material is cued up by hand, and CDs are played directly into the Avid without any thought about recreating the job or, worse, starting over again if disaster were to strike. If the time is available, then seriously consider dubbing all nontimecoded material to a timecoded, high-quality format source. The potential slight loss of quality involved in dubbing will be the difference between quickly recreating what you have done or matching things by eye and ear. If you cannot timecode all your sources, then seriously consider copying to a large inexpensive FireWire or USB drive for the evening after you have finished digitizing all your media. Remember, a digital nonlinear project is never done, you just run out of time, money, or both.

Conclusion

Beginners don't grasp these techniques from the entry-level course. The techniques are usually a combination of changing some of your work methods and taking advantage of some unfamiliar functions. There are many other keys, modifiers, and tips and techniques in the Avid editing systems, but these are the main areas to concentrate on. Use the keyboard more and spend more creative time using the Trim mode. Start to think nonlinearly about navigation and the structure of the sequence. You will find your speed incrementing in leaps and bounds.

4

AVID ADMINISTRATION

"It's hard to be fully creative without structure and constraint. Try to paint without a canvas."

—David Allen

Some people are lucky enough to have their Avid system administered by someone else—if you're one of them, hand them this chapter and go back to cutting. However, the day-to-day reality is that most people are responsible, at some level, for their own system. After all, if you have created a difficult situation for yourself, it is you who cannot go home until it is resolved. A smooth-running session makes you look good, period.

This chapter explores the peripheral issues of owning and maintaining a professional editing system. This includes environment, media management, and networks. For the Avid administrator or postproduction supervisor, getting media in and out of the system is as important as anything you do with it while editing.

Room Design

It is easier to administer a system that is set up correctly and put into a room that is well designed for the Avid system. Fewer constraints are placed on the design of a digital, nonlinear suite than a traditional, tape-based suite, but this doesn't mean you can just plunk the equipment down in a pile! There is flexibility to move all the equipment into another room and keep the bare minimum in the suite or move the noisiest parts, like drives. There are technical implications to making cables longer, as you will see. Suites need to be designed so they can be serviced and simple maintenance can be done without disruption. If you plan to use the suite as an online finishing facility, then you should treat it like an online suite. Use an external waveform and vectorscope, high-quality speakers, a high-quality third monitor, and seriously consider a patch bay. There is nothing worse at 2 AM than hauling video decks around so that you can do two dubs. If connections are made through a professionally installed patch bay, there is a better chance the dubs will come out well!

Probably the biggest consideration after editor ergonomics is the noise level. Many people make the mistake of putting all their equipment in one room. Though it is simple, it can make for a noisy environment. The drives and the computer processing unit (CPU) are likely to be the noisiest elements of your system because of their fans, so they are the most important to move to another room.

The slickest installations I have seen have completely relocated the CPU and the drives to another room. You'll need to use a KVM (keyboard, video, mouse) extender to connect your keyboard mouse and monitors to the relocated equipment—these are easily found nowadays even for dual-DVI (digital visual interface) monitor configurations. Of course, if you have no central machine room and the editor must load tapes and capture, there should be a connection for a video deck's video, audio, and control cables. If all the hardware is in another room, there should be telephone links between the two rooms (preferably with speakerphones for troubleshooting purposes). The best facilities have gone back to the central machine room design after the initial years of isolated Avid suites. It is important to have a range of decks during the course of an edit, so multiple decks need to be available. Tying up a deck all day when it is not really needed is just as bad. A central machine room also makes it easier to get to the equipment for support or upgrading. There is nothing worse than working under a table with a flashlight in your mouth, the telephone in one hand, and a screwdriver in the other. Unfortunately, this happens way too often because of thoughtless room design.

If you must have all of your equipment in the room you're editing in, then look into buying a sound-isolating rack. They typically come in desk-side half-height configurations, but full-height configurations are available as well. You can usually find them for sale at sound equipment supply sites. These racks typically include airflow controls but they'll only keep your equipment cool and happy if you install them correctly. *Remember:* Those cable snake holes are there for a reason. Don't just leave the back of the rack open—not only will you dramatically increase the equipment noise, but you're increasing the risk of your equipment overheating and failing.

One of the nicest things about many of the Avid suites I have worked in is the addition of windows. There is nothing better to clear your head than to stick it out a window for a few minutes, and having natural light is a nice change; however, I must fall back on the admonition that if you are doing color correction or shot matching of any kind, you need to have complete control of the lighting. (We talk more about a proper color-correction environment in Chapter 10.) Pick the neutral color temperature of the lighting carefully and by all means avoid fluorescent fixtures. Color temperature of sunlight changes during the day, so a shot captured and evaluated in the morning may not look like those adjusted at noon unless the ambient lighting is indirect and consistent. If you have windows, make sure you can pull the room-darkening shades and keep glare off the monitors.

If you also are planning to do final sound mixing, make sure the room has been deadened. Apply sound-absorbing foam around the room or make sure that the room has enough carpeting and wall hangings. Mixing in an empty office is probably a bad idea. Investing in good speakers and a real amplifier pays off quite quickly. You may also want to consider cheap speakers with an A/B switch to the reference monitors or pumping the Avid audio output through a standard home television monitor in the suite. There is nothing worse than listening to your wonderful sound mix at home and having it sound muddy from overpowered bass.

Personally, I feel no suite is complete without two phones, a trash can, a box of tissues, and a dictionary. You'd be surprised where people set up these temporary suites: attics, basements, bedrooms, storage closets, hotel rooms, boats, bank vaults, Chinese laundries, and ski lodges. Try to minimize any environmental impact like heat, dust, or jarring motion (like editing in the back of a moving truck). Even 5–10 degrees of difference in temperature can add or subtract useful years from the life of the equipment.

Electrical Power

The final and probably most important piece of equipment you need under any conditions is an uninterruptible power source (UPS). If you are running a system now without a UPS, you should nonchalantly put down this book and run to the phone to order one now. You are living on borrowed time. A UPS regulates your power, giving you more when your electric company browns out or less when you get a spike. If power fails completely, a UPS gives you enough battery time to shut down the system in an orderly fashion and avoid crashing, losing work, and potentially corrupting important media or sequences. A UPS makes your equipment run with fewer problems, and you will be able to charge for more productive hours of use.

The real question is not whether you have a UPS, but how much of one do you need? They are figured in the confusing scale of volt-amps. The math is not so hard if you can find out what each piece of equipment needs for electrical power for either volts or amps. The numbers are usually in the manuals or on the equipment itself. There is information on the Avid website under the Customer Support Knowledge Center that lists the power requirements of each piece of equipment. But don't neglect connecting the tape deck. What happens to your camera's original tape if the power goes out when you are rewinding?

Here is how to figure the size of an uninterruptible power source that is sold in volt-amp models:

$$\text{volts} \times \text{amps} \times \text{power factor} = \text{volt-amps} \times \text{power factor}$$
$$= \text{watts}$$

The power factor for computers is between 0.6 and 0.7, so you can look at the same equation as:

$$\text{watts} \times 1.4 = \text{volt-amps for computers}$$

And because you don't really trust manufacturer specs for the UPS and you buy more than you need to accommodate future expansion, add another 33 percent on top of what they recommend.

Keep in mind that before you get a true blackout, you will probably suffer from sags and brownouts. These may cause the UPS to use up some of its battery power to keep you going until the "big one" hits. Then, when the power comes back on, you can count on a serious power spike. A spike can cause damage to boards, RAM, and drives, and that damage may not show up until days, weeks, or months later as the parts start to fail prematurely. The fact that all power is going through a series of batteries and power conditioners with a UPS before it gets to your delicate equipment should give you a warm feeling in your stomach. Just make sure when you get it all hooked up that the battery is actually connected, since some UPS manufacturers ship the equipment that way.

If there is any question about what to put on the UPS, imagine using that device full tilt when the power goes out. Ever see a one-inch machine lose power while rewinding a finished master tape? Not good. A cassette-based tape deck will almost certainly crease the source tape if it is rewinding when the power goes out.

Even if all the lights and your monitors go out, you can always save and shut down quickly using just the keyboard. In an emergency, remember:

- Ctrl/Command+9 activates the Project window. You want to save the whole project, not just the active bin.

- Ctrl/Command+S saves everything.
- Ctrl/Command+Q quits the application in an orderly way. Quitting will save everything first, but trying to save the project should be your first step anyway. You may have to hard boot the system to get control back after a power hit. Saving should be an automatic first step in any emergency procedure.
- Enter to confirm that you really do want to quit.

This sequence of keystrokes avoids the chance of corrupting the project from being shut down improperly and can be performed (if you really have to) with the monitors blacked out. A UPS has saved me literally a dozen times—I even use one at home on my computer and NAS (network attached storage). After your first serious power hit, what is the real cost of replacing your system?

Ergonomics

Human ergonomics has been written about at length in other places, so just a quick word about it here. Don't scrimp on chairs. They make the difference between happy editors and editors in pain. Get chairs that can adjust armrests, back, and height. Many people swear by armrests, and with a keyboard and wrist rest at about the same level, there is less chance of wrist strain. Keyboards can be put on sliding shelves below the workspace. Keep the back of the hand parallel to the forearm to reduce wrist strain. The relationship of chair, keyboard, and monitor cannot be underestimated as important to the creative process. Some editors even cut standing up, just like they did when they cut on a Moviola!

Media Storage and Management

Let's discuss the nuts and bolts, the bits and bytes, of what happens when you put media on your system. The Avid editing application is an object-oriented program, which means that many things you do create an object. Capturing media and rendering, importing, and creating sequences and bins all create different kinds of objects. It is the relationship between those objects that allows you to combine things in such interesting ways.

The editing system sees only the media files that are on the media drives in the folder named OMFI MediaFiles (for OMF media) or Avid MediaFiles (for MXF media) depending on the type of media you are using. This folder is created automatically and named by the software when the application is first launched and the drive is used for capturing. Both of these folders must stay on the root level of a drive or partition and cannot be renamed. If you do, the media files inside the folder go offline and are no longer accessible to you when you try to edit.

While media files are stored directly inside the OMFI MediaFiles folder, media files in the Avid MediaFiles folder are stored inside a series of subfolders. If you open the Avid MediaFiles folder you'll see a folder named "MXF," and if you open that folder you'll see one or more folders, each with an associated number (at the very least there is a single folder named "1"). The reason for this hierarchy is that the Avid MediaFiles folder structure is designed to store more than one type of media. Currently, only MXF media is stored, hence the "MXF" folder, but it is entirely possible that future versions of Avid editor will store other types of media in their own associated folder.

Now, about those numbered folders in the Avid MediaFiles/MXF folders. These folders are designed primarily to help reduce the number of individual files in any given folder. Though modern operating systems can now handle thousands and thousands of files in a given folder, that wasn't always the case. And even though they *can* handle thousands and thousands of files in a folder doesn't necessarily mean that it is a good idea! In fact, you'll find that the Avid system will automatically generate a second folder named "2" once you reach about 10,000 files in folder "1."

You don't have to just let the system manage your MXF media, though. You can use a folder numbering scheme to keep one set of media (e.g., originally captured media) from another (such as rendered media). The Avid system will always write new material to a folder numbered "1" until it determines that folder is full. But it will read media from any numbered folder it finds in the MXF directory.

Give this a try sometime: Quit Media Composer then open up an Avid MediaFiles folder on one of your storage drives or partitions and then open the MXF folder inside that. Renumber that folder. I recommend that you only use numbers because alphanumeric names are not officially supported, but I've typically found that if you always start with a number you can usually name the folder anything you want. This technique is especially useful when working with P2 media as it lets you keep your P2 media organized.

OMF or MXF?

As I mentioned previously, Avid supports both OMF and MXF media. And by both I mean that you can freely mix and match OMF and MXF media in a project or sequence. You can even mix them in a single master clip as it is possible for your video media to be MXF and your audio media to be OMF or vice versa. Many folks use OMF media because that is the older type of media and a type they've been using for years. For the most part, the two types of media are interchangeable. But MXF has some distinct advantages over OMF when it comes to storing video media:

- OMF files are limited to 2 GB in total duration. The Avid system can be configured to automatically span multiple OMF files during capture, but this 2-GB limit can be a real pain when you're trying to work with high-resolution material. MXF media files can grow much larger.
- OMF only supports 8-bit media while MXF supports both 8- and 10-bit video media.
- OMF can only store NTSC and PAL media. MXF can also store 720- and 1080-line HD media and beyond.

Regarding audio, both OMF and MXF are uncompressed and both support all of the expected sample rates and bit depths. MXF also supports compressed audio though you'll likely only encounter this when working with XDCAM proxies or other proxy material.

So which one should you use? Often the facility you work at, especially their audio postdepartment, will define the format for the audio media. And the postfacility may have a preference for the format of your standard-definition (SD) material. But if you're working with high-definition (HD) material, you have to use MXF.

Compression, Complexity, and Storage Estimates

None of this clever manipulation of objects solves the basic problem of running out of space on the media drives. It is only a matter of time before this problem occurs, and you should prepare for it in an organized way. There are several ways to tell when you are going to run out of space. The Hardware Tool under the Tools menu (also under the Info tab of the Project window) gives you a bar graph of how full the drives are relative to each other, the amount of storage empty and used, and percentages full (if you have Tool Tips turned on). This is good for figuring out where to start capturing the next job, but it does not give you the amount of space in terms of amount of footage.

If you need precise numbers, then open the Capture Tool and choose the tracks you think will be needed the most (this may be called the Digitize or Record Tool on your system, but all new Avid systems have moved to use the term Capture). If you are working with material that has sync audio, then turn on all the tracks. But if you are working with mostly MOS (silent) film transfer that will be cut to an existing soundtrack, get the estimate with only the video track turned on. Make sure the compression level or resolution you will be using is set correctly. Video takes a massive amount of space to store, even compressed, compared to audio files. The Capture Tool gives you an estimate of how much time you have on each drive.

Table 4.1 provides you with some consumption rates for popular Avid resolutions. Simply multiply the number of minutes of footage you have to capture by the consumption per minute to estimate the storage required. Note that audio takes up such a small amount of space as compared to video that it isn't typically that significant (audio consumes roughly 1250 MB per channel per hour).

Table 4.1 Resolution Storage Requirements

Format	Resolution	Storage Consumption
NTSC or PAL	1:1	1.22 GB/min
	2:1	526 MB/min
	3:1	345 MB/min
	10:1	132 MB/min
	15:1s	31 MB/min
1080i/59.94	1:1	8.68 GB/min
	DNxHD 220	1.54 GB/min
	DNxHD 145	1 GB/min
1080i/50	1:1	5.79 GB/min
	DNxHD 185	1.28 GB/min
	DNxHD 120	0.85 MB/min
1080p/23.976	1:1	5.56 GB/min
	DNxHD 175	1.22 GB/min
	DNxHD	0.81 GB/min

MB = megabytes; GB = gigabytes.

If you select all the clips in a bin and choose Ctrl/Command+I (Get Info), the Console opens and gives you a total length of all your clips. This is a powerful way to see whether you have enough space on your drives to recapture everything.

Unlike video or film, compressed images are judged by their complexity. A complex image takes exactly the same amount of space to record on videotape as a simple one! When you capture an analog image to disk or "ingest" an already digital image, the level of image complexity is important. The more information and detail in the frame, the more space it takes to store and the more difficult it is to play back. Playback from a disk-based system is a question of throughput or how much information can be read from the hard drive, pushed through the connecting buses into the host memory, and out to the monitors or tape decks. This is why a slower system may not be able to handle high-resolution images—it cannot get the information from the drives fast enough to play all the information in real time with effects and audio.

There are three categories for captured images: *uncompressed, lossless compression*, and *lossy compression*. All compressed images on Avid systems are lossy, where redundant information is thrown away during capturing. As you move closer and closer to uncompressed quality, you pay a higher cost for hardware and disk space. You must carry over every pixel of every frame, no matter how redundant that pixel is. Uncompressed images demand faster computers, wider bandwidth, and much more disk space on faster, striped drives.

Lossless compression is many times touted as better than uncompressed (or noncompressed, as some insist) because it takes less disk space. Lossless compression is associated more with programs like WinZip™ for compressing documents before posting them on the Web for downloading. The difference when compressing something variable, like a moving video shot, is that as the image gets more complex, the compression is less effective. Potentially, under a wide range of circumstances, a lossless compressed image could be larger than the equivalent uncompressed image (compression information and the less compressible image are added together). If the editing system is designed to take advantage of a low bandwidth as a benefit of smaller file sizes, you may have some playback problems. The system may impose a rollback, where a maximum frame size is imposed by throwing away information (lossy) when the frame size gets too big. If they don't do this, they must prepare to handle even larger frame sizes than the uncompressed system and lose much of the benefit of lossless compression.

The reality is that compression is here to stay. Though high-end postproduction is still done with uncompressed images, even in HD, virtually all methods of transmission and distribution use compression. And virtually every digital video acquisition format today is also compressed. One primary main reason to use uncompressed images is that the less compression artifacts in your image, the less likely those artifacts will compound into worse artifacts in the final product. Uncompressed (or lightly compressed) is also useful for archival purposes since there may be a future compression method that works better if it is starting with no compression at all. These uncompressed SD images will most likely be upconverted to HD at some point in the near future, and any compression may be visible in context at that time.

Deletion of Precomputes

Predicting available space on media drives must go hand-in-hand with keeping track of rendered effects, or precomputes. Imported graphics and animation also take up space. Every time you render an effect, it creates a media file on the drive. Even though you may cause that effect to become unrendered or delete it from the sequence, that file still lurks on your media drive. This is actually for a very good reason, for both undoing and for all the multiple versions of that sequence, but it means you need to pay attention to how full a drive has become even though no one has captured to it that day.

Deletion of precomputes is one of the most important things an Avid administrator can oversee. One of the most common calls to Avid customer support is when, during a session, a system grinds to a halt because it is too full of thousands of tiny rendered effects. Are all those effects necessary? Probably not, and now the editor or the assistant must be walked through the process of deciding what can go. One of your most important responsibilities in making sure that sessions start and end smoothly is to keep an eye on how many precomputes are on the system and how many are really important. The Avid editing systems do not keep track of how many sequences are created during a project. There could be multiple CPUs accessing the same media or archived sequences that are modified on another system and brought back. Since the system is so flexible, there is no way the system could definitively know the number of sequences created. What if the system was to make very important decisions for you, like deleting "unneeded" rendered effects? What would happen if you called up a sequence you had spent hours rendering to find that the software had neatly deleted that media file automatically, thinking you were done with it? Just because you deleted the effects sequence from sequence version 15 doesn't mean you don't want effects on sequence versions 1–14! There are too many variables, and this decision is too important to leave up to an automated function at this stage in the technology.

That being said, there is, in fact, some auto-deletion of precomputes going on under your very nose! But, as it should be with all automatic functions that cause you to lose things, it is very conservative and you may not even notice. The only auto-deletion of precomputes occurs when you are making creative decisions quickly and removing or changing effects. If you are rendering the effects one at a time and then quickly deleting them, there is at no time any opportunity for the system to save the sequence with those effects in place. There is no record that they will be needed in the future because they have not been saved. Saving happens automatically at regular intervals and, when you render effects in a bunch, they are saved as the last step before allowing you to continue. Every time you close a bin, you also force a save of the contents. That is why the software auto-deletes precomputes only when you are rendering one at a time and quickly removing or unrendering them by changing and tweaking. If a save occurs while an effect is in a sequence, the precompute is not deleted automatically.

Every little bit helps to keep the drives unclogged, but you still must evaluate the amount of precomputes and delete them on a regular basis. This is really not so hard even though it is a little intimidating at first because it involves deleting material that someone (probably the editor) may need if you get it wrong. This chapter will deal with the isolation of precomputes when we look at efficient deleting strategies.

It is certainly a good idea to remember to delete all of the pre-computes when you finish your project. I know far too many editors who have remembered to delete their captured video and audio media, but forgot about the precomputes! If you don't delete them they won't expire on their own, and I guarantee you that they will eventually fill up a drive! (I remember an unscrupulous rental agent who, when faced with a drive filled with pre-computes, told the production that "these things happen" and they'd need to rent more drives.)

The Importance of Empty Space

Remember the media database? Any file that keeps track of all the media is a concern if you overfill your drives. That file must be allowed to enlarge to deal with the many files you add during the course of editing. If there is just not enough room for the media database file to update and grow larger, you may have media file corruption and eventually drive failure. A good rule of thumb is to leave at least 5–10 percent of each partition empty. This limit is flexible, but if you detect slow performance in the form of more underruns or dropped frames, then consider moving media to an emptier drive. It is also a good idea to erase media drives completely after a job has been completed. Don't get initializing a drive confused with low-level reformatting! That is only a very last resort to save a dying drive. However, reformatting causes a drive to lock out any bad sectors—those same bad sectors that may have been giving you problems will be eliminated from future problems.

Consolidate

The best tool you have for moving media while inside the software is Consolidate. Consolidate can be used for two main purposes:

- Moving media from multiple drives to one drive.
- Eliminating unneeded material.

Take whatever you need to move, either the media relating to an entire bin or a finished sequence, and consolidate it to another drive.

There are some choices when you consolidate that may make things less confusing. First, you need enough free space on your drives equal to the amount of material you wish to consolidate. You are able to specify a number of drives for consolidation. Using the list of drives means that, even if you run out of space on one drive, the next drive in line will take the overflow material. The second drive will take the material until it is almost filled, then the third drive, and so on, until the sequence is finished consolidating.

These are the two major reasons to use the Consolidate function, although as a quick troubleshooting tip you may choose to consolidate a clip that is not playing back correctly. If the clip plays back better after consolidating it to another drive or partition, you may have a drive problem or you may have captured the clip to a drive that was too slow to play back that resolution.

Consolidating a Sequence

The beauty of Consolidate is the advanced way that it looks at everything that is needed in a sequence and copies only that. There are, after all, other ways to copy media, but it is very difficult to tell at the operating system level what clips are really necessary to play a sequence. The Consolidate function will search all your drives for you, gather only the bits you need, and then copy them to the desired drive.

Consolidating will break the sequence into new individual master clips and copy just the material required for the sequence to play. This creates shorter versions of original master clips because you are copying only the bit that is needed. You can then selectively delete unused media.

Consolidate is especially important at the end of a project when the final sequence has been completed and it is time to back up the material. Instead of backing up all the media, you consolidate first and back up only the amount of media that was actually used.

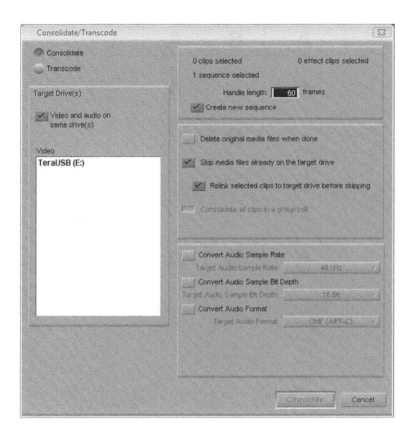

It is also useful before sending audio to be sweetened on a digital audio workstation. Delete the video track from a copy of the finished sequence, then consolidate this sequence and make sure the audio media files go to a removable drive.

After you select the sequence and consolidate it, the system looks at the original master clips to determine exactly what is necessary. If you have an original master clip that is five minutes long, but you used only ten seconds of it, then only that ten seconds will be copied. The new ten-second master clip will have the original name with ".new.01" if it is the first time in the sequence that shot is used. If you use another five seconds of the same original master clip, then you will have a second consolidated clip with ".new.02," and so forth.

The Consolidate process also allows you to specify handles, or a little extra at the beginning and end of the clip, so you can make some little trims or add a short dissolve later. If the handles are

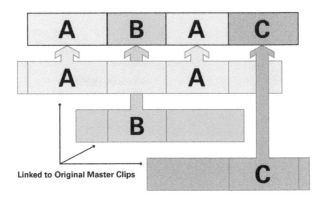

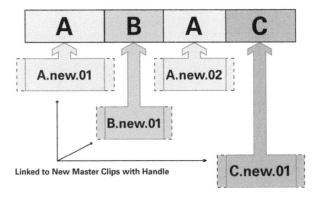

too long, you will throw off your space calculations, so be careful; however, if the handles cause two clips to overlap material, then Consolidate will combine the two clips into one new clip rather than copy the media twice.

Consolidating Master Clips

If you want to move media from multiple drives to a single drive, you can consolidate master clips. This will take all of the media linked to the clips in a bin and copy them in their entirety to another drive. Notice the difference between move and copy. You are really just copying and then must decide whether or not to delete the original. Copy and delete are the two steps that make up the move.

Consolidating master clips is a fantastic method for being able to clear off multiple drives and put everything from one project onto a single drive or onto a series of drives so you can remove them or back them up. Consolidating master clips does not shorten material. It is just convenient and takes a complicated media management task and makes it one step. Be careful not to overfill any one partition.

If the idea is to move media from multiple drives to one target drive, then you must ask yourself, "Have I used this target drive before for capturing this project or a previous consolidation?" If the answer is yes, then you may already have some of the material you need on the destination drive. You don't need to copy the media twice! Be sure to check the option "Skip media files already on the target disk." You would use this only if you were consolidating master clips and not sequences. Using the media already on the drives is perfectly fine. When the Consolidate function finds the long, original media file already on the target drive, it will leave it alone, untouched. This is the best setting for moving media from multiple drives to a single selected drive.

There is a secondary choice that becomes available to you only when you check "Skip media files already on the target disk," and that is the somewhat confusing "Relink selected clips to target disk before skipping." You want to make sure that when you skip the media files that are already on the drive that the consolidated master clips are linked to them. When the system makes consolidated master clips that link to existing media it doesn't add a ".new" to the end of the master clip name. It will add ".old" to the master clip in the original bin, however, since it needs to distinguish between the two clips.

Consolidate Summary

If you are consolidating sequences:

- Choose the drive(s).
- Determine the handle length.
- Uncheck "Skip media files already on the target drive."

If you want to delete the original media after you have created the shorter consolidated clips, then check "Delete original media files when done." If you are feeling prudent, go back and delete the original media in the Media Tool after consolidation. This is especially important if you are consolidating multiple sequences and they all share media. Do not delete the original files until you have consolidated all the sequences that use the media! If you are working with multicam, you should choose to consolidate all the clips in the group if you want to continue to have all the camera angles available in the consolidated sequence.

If you are consolidating master clips:

- Choose the drive(s).
- Choose "Skip media files already on the target drive."
- Choose "Relink selected clips to target drive before skipping." Do not choose this if you are continuing to work on the project and someone else will take the consolidated media away and work on them elsewhere.

You probably do not want to delete original media here and will proceed with more sophisticated media management later. If you do not delete your media now, you will have a mix of ".old" with original clips on the original drive and ".new" with the new clips on the target drive.

Fixing Capture Mistakes with Consolidate

Let's take the example where you have captured too many tracks. Perhaps you (or someone just like you) weren't paying attention when you were capturing and you captured a video track with a voiceover master clip. Delete the unneeded video and keep the audio through Consolidate. You can take this master clip and make a subclip of the entire length, making sure that you turn off the tracks that you don't want. If you have captured voiceover and accidentally captured (black) video, turn off the video track while it is in the Source window before you subclip the master clip. The subclip will be audio only. Highlight that audio subclip in the bin and choose "Clip > Consolidate." The Consolidate function will copy only the media you want to keep. You will have a new subclip and a new master clip with only the audio tracks and can delete the original master clip to free up disk space.

There is even a faster method to do this particular technique. You can find the clip in the Media Tool and delete just the media file that you don't need. You are given a choice of whether to delete the audio or video media, so for the video with voiceover problems, pick the video media for deletion. Then unlink the clip and choose "Modify." Change the master clip to reflect that it is now just audio. You will not be able to change the number of tracks of the master clip unless you unlink it first, then relink the clip back to the audio media. This method also ensures that you don't end up with strange media management problems down the line when batch capturing or restoring from an archive.

Subclipping Strategy with Capture

You can also use the subclipping then consolidating method just to shorten your master clips after you decide what part of them you really need for the project. In fact, many people like this method as a general strategy and capture master clips that are quite long, maybe an entire scene or the full length of an already edited master tape. This creates fewer files on your drives for the computer to keep track of, creates fewer objects for an object-oriented program, and can speed up performance. Then subclip all the sections you will actually use, consolidate the subclips to create new master clips, and delete the rest.

Using the Operating System for Copying

If your goal is to move an entire project to another drive, you may be better off working at the desktop level. If you have been using MediaMover (www.randomvideo.com), your job is a snap. MediaMover will search all your media drives, find all the media from a specific project, and then move that media into a folder with the project name. Copy the entire folder with the name of your project from each of the affected media drives. It is easy to do a Find File (Windows/Command+F) and copy every folder found with the project name even if you have a dozen drives. If you are backing up or moving media around and you don't have MediaMover, then buy MediaMover. It is as simple as that.

The situation may be complicated if there are two different resolutions to keep track of in this project. You may want to copy only the low-resolution material and leave the high-resolution material on the drives or vice versa. Planning helps here and, if you are on a Macintosh, I recommend that you use the Labels function provided by the Macintosh operating system (this is for OS 9.x only). If you go to the Control Panel for labels, you can change what the colors represent. Change the blue label to

"Project X Low Res" and the green label to "Project X High Res." Select all media files immediately after you have captured them and change their label color. You can sort the media files by label color and copy or delete only the files you want. This also allows you to track down those stray media files that always manage to escape even the most careful herding.

Whether you choose to use Consolidate or Copy on the Finder level, you need to keep track of all media. It must be searched for, backed up, copied, or deleted. The number of objects on the system should be checked periodically, and unneeded rendered effects or precomputes must be deleted on a routine basis. If your drives fill up, the session stops.

Deciding What to Delete

A process in which an administrator or an assistant must be very careful and yet very efficient is with the deletion of unneeded material. Sometimes this can be agreed upon mutually with the editor and you can eliminate all of the material for Show 1 when you are well under way with Show 2. Many times different projects share the same material and you must be careful not to delete that which is needed by both. The Media Tool and MediaMover can both be used efficiently project by project, but this may not be good enough. You must find some other criteria to sort or sift by, protect certain shots, or change the project name of the material you want to keep.

Using Creation Date

Creation Date becomes a very important criterion to look for individual shots, and it is often overlooked by many assistants. It does not work like the modified date on the desktop, which updates to reflect the last time someone opened and modified the file. The creation date is stamped on the clip when it is logged, so if the shot is logged and captured on the same day, this becomes a useful heading to eliminate material that was captured at the beginning of a project.

I especially like to use Creation Date as a heading in my sequence bin. I duplicate my sequence whenever I am at a major turning point or even if I am going to step away for lunch, dinner, or a snack. When I duplicate the sequence, Creation Date time stamps it so I know that I am working on the latest version. Then I can take the older sequence and put it away in an archived sequence bin (or several archive bins as the bins get too big). I have control over the exact times that I have stored a version, instead of leaving it up to the auto-save, and I can keep the amount of sequences in any open bin to a minimum.

Sequences will be the largest files you will work with. In an effort to have the bins open and close quickly and not use up too much RAM, I try to keep the bins small and keep only the latest version of any sequence.

If you are trying to determine which sequence is the latest version and the editor is not present, Creation Date is the best tool. There is always the chance that the editor duplicated the sequence and continued to work on the old one, but you should have an agreement with the editor about how to determine this crucial fact.

Using Custom Columns

There are many other criteria you can use for sorting and sifting. If you plan it well, you can create custom columns to give you an extra tool to work with. Some people will create a custom column with an X or some other marker to show whether a shot has been used. One way to use sifting powers of the bin is to change the sift criteria to "match exactly" and have it search for a blank space in a custom column. This way, any shot that has not been marked is called up. The Media Tool is the only way to search for media files that are online across bins. You can create a Media Management bin view, which can be used in the Media Tool. Your custom columns will show up there, too.

Basic Media Deletion Using Media Relatives

Another way to find out if a shot has been used within a project or sequence is to use the Find Media Relatives menu choice that is in every bin. This is the best way to search across bins to clear unneeded shots, other than consolidating and deleting the old media.

The most useful way to use Find Media Relatives is with sequences:

1. Open the Media Tool. You have a choice whenever you open the Media Tool to show master clips, precomputes, or individual media files for all the projects on the drive or just for this project.
2. Show master clips and precomputes for this step. I like to show "All Projects" when checking for media relatives because many times the bins I am working with came from another project.

3. Open all your relevant sequence bins for this method to be accurate.
4. Put all the sequences that you want to work with into one bin.
5. Select all the sequences and then choose "Find Media Relatives" from that sequence bin. The system searches all open bins and the Media Tool, and highlights all the master clips, subclips, and precomputes that you will need to keep.
6. Go back to the Media Tool and choose "Reverse Selection" in the pulldown menu.
7. After reversing the selection to highlight all the unused media, press Delete.

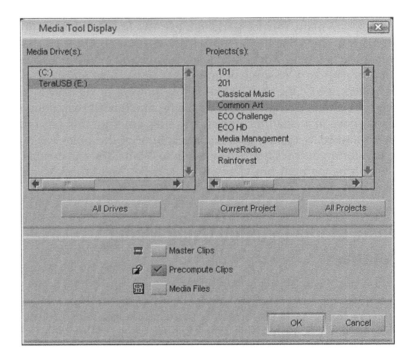

Keep in mind that Find Media Relatives does not unhighlight something, so before you begin this operation, make sure nothing in the Media Tool is already selected. Also, give everything selected for deletion one final look. This is the last time you will see these files, and you don't want to trust anything or anyone except yourself at this stage—there is no undo. Don't do this when you are tired! Not all of us are "morning" people!

You can choose to show only precomputes in the Media Tool. You will see all the rendered effects for a project with all the effects on your drives. You can use the previous method, Find

Media Relatives, to track down the precomputes used for the sequences you've selected and then, remembering to reverse the selection, delete the unneeded precomputes. This process can be done every day if you are working on intensive effects sequences, like in a promotions department. It doesn't need to be done unless you need the space or memory requirements are climbing into the hundreds of thousands of objects. There is always the chance you might delete something you want to keep, so be careful.

Occasionally you will want to find exactly where a specific file is and for some reason, Consolidate, MediaMover, or the Media Tool is not sufficient to perform the task you need. You can use a function called Reveal File, which will go to the desktop level and highlight the media file associated with a specific master clip.

Lock Items in Bin

If I have achieved my goal of making you slightly paranoid about deleting material during a project, then you will approach this task with the proper amount of stress. There is a great way to relieve some of this stress that does not require a doctor's prescription. Many years ago I spent a week at a major network broadcast facility observing their operations. They needed someone who knew nothing about the individual projects being edited to come in late in the day and clear drive space for the next morning. On longer jobs, this person would be familiar with the exact needs of the editor. In this case, they were working on as many as five or more projects a day. At the end of the day, some material needed to stay—material that had been grabbed from their router and therefore had no timecode—but the majority of the material needed to be deleted. This is where the idea for Lock Items in Bin originated.

It was supposed to work like this: As the editors gather shots and use them, they decide that they need some shots for the continuing story tomorrow and clone the master clips (not the media). They Option-drag and clone the master clips into a bin named Stock Footage. At intervals during the day, the editor goes to the stock footage bin and selects all (Ctrl/Cmd+A). The editor chooses "Lock Bin Selection" from the Clip pulldown menu (or right-mouse/Shift-Ctrl+click). A lock icon appears in the Lock heading that is displayed as part of the view for that bin. The file is locked automatically on the desktop level so that if some intrepid assistant decides to throw everything away, he or she sees a warning that certain items are locked and cannot be thrown away. Any good assistant, however, knows that on the

Macintosh if you hold down the Option key when emptying the trash, you can throw almost anything away and cheerfully ignore these pesky warnings. On Windows, you need to go to the file, right-click the icon, and unlock the file under Properties. You can select many files simultaneously for locking or unlocking, but if you drag them to the Recycle Bin you will get a warning that the file is read only. It may allow you to delete it at that point, but since you are going to the Recycle Bin you will get the chance to retrieve it. You can see whether a media file is locked on Windows by looking at the Properties to see if the file is read only.

A happy side effect of the Lock Items in Bin command is that it can also be applied to sequences. Subsequent copies of a sequence are also locked automatically when the sequence is duplicated. This helps avoid the all-too-common problem when people have mixed clips and sequences in the same bin and they decide to delete files. They select all the files, delete them, and then, "Hey, where did my sequence go?" Certain types of objects are selected automatically as choices to be deleted when you select all files in a bin. The sequence is automatically not checked for deletion if it is mixed in a bin with clips. If a bin has only sequences then they will indeed all be checked for deletion. Deleting a sequence by accident is usually someone's first visit to the attic, only to recall the version of the sequence he or she worked on 15 minutes ago!

There are ways to outsmart the Lock Items in Bin command if you are determined. You can duplicate the locked clip and then unlock the duplicate. Because both clips link to the same media, you can delete the media when you delete the second clip. You can also unlock the media at the desktop level and throw them away without even opening the editing application. But you can't delete the media until someone, somewhere, unlocks them. Using these methods is really the honor system for media management.

Changing the Media's Project Association

There is another way to organize shots for easy media management: change a master clip's project name. There can be confusion about exactly which project is associated with which media. It is so easy to borrow clips from another project that you may not even know that certain shots are from another project unless you choose to show that information in your bin headings. This is why I create a custom bin view just for media management. It shows the project name, lock status, creation date, tracks, video resolution, disk, and any other customized headings that relate to media management. If the clips have been borrowed from another project and brought into the new project, even if you recapture them, they retain their connection to the project in which they were logged.

Project affiliation makes a big difference when it comes to recapturing at a higher resolution. Because all the shots from one tape are captured at once, if you are trying to recapture from Tape 001 in Project X, and the clip is really from Project Y, the software considers these two different tapes. While batch capturing, the system will ask for Tape 001 twice since it really should be two separate tapes if it was logged correctly. There are several steps to changing project affiliation. The first thing to keep in mind, as mentioned in Chapter 2, is that the project name is part of the tape name. The trick to changing the project name is to change the tape name to a tape name from the right project. All tapes have a Project column in the Select Tapes window when assigning tape names. This allows you to see which project each tape is from.

In Capture mode, whenever you show the list of tapes that have been captured and you check Show other projects' tapes, you might see multiple "Tape 001" listed from other projects. Use the "Scan for Tapes" button in recent versions to make sure the system has updated all the tapes captured and used on the system.

Here is the best way to change the project name of a master clip:

1. Open the new project.
2. Open the Media Tool and show the original project media files.
3. Consolidate the master clips into a new bin in the new project.
4. Do not skip files that are already on the drive.
5. Select all clips in the bin and choose "Modify" from the Clip menu.

If the tape name you want has not been used before in this project, then you can create a new tape name. You can choose an existing name if you made an earlier mistake and need to rename these clips so they all come from an existing tape in the new project. As you finish with the modification process, you get a series of warnings that are important if you are working with key numbers in a film project since the new clips need them reentered after you modify the tape name. A film project will not allow you to change a source name of a clip without unlinking first. If you are not using key numbers, just check OK.

Now the clips are associated with the new project as far as the Media Tool and the headings in the bin are concerned. The change has not really affected the actual media file at this stage, just the master clip in the project. This is not enough for MediaMover to recognize that there has been a change because it looks only at the media file and not the project. The project information about a media file is actually recorded in the media object identification (MOB ID—it has nothing to do with gangsters). The MOB ID is attached to the media file when it is captured, and the media file itself must now be changed to update this information. Here is another use for Consolidate: You need the space on the drives to copy the files you want to change.

To change the project name of the media files, select all the clips and Consolidate with the following options checked:

- Have the old name link to the new media files.
- Do not skip over the media files if they are already on the drive.
- Delete the old media files when you are done.

If you want two sets of the media—one set of media in one project and one set in another—do not delete the original media files after you consolidate.

Relinking

All master clips, subclips, sequences, and graphics must link to media in order to play. When you log a shot into a bin, you create a text file. When you capture the shot, you create a media file on the media drive. The relationship between the master clip and the media file is considered to be a link. If the master clip becomes unlinked from the media file, it is considered offline. The media may be on the media drive, but if the master clip is unlinked, the media are considered offline and unavailable. In order to link, you must highlight the clip and choose "Relink" from the Clip pull-down menu. The ability to link, relink, and unlink is very important to sophisticated media management. We will revisit these concepts over and over again in this book. Following are eight rules that govern this complex and confusing behavior.

Rule 1—Tape Name and Timecode

All linking between a master clip and a media file is based on an identical tape name and timecode with media in the OMFI MediaFiles or Avid MediaFiles folders. You can also relink by key number for picture only, which is excellent for linking media from a new film transfer to a finished sequence. There will also be a choice in all models to relink based on resolution. This will allow you to force master clips and sequences to link to media based on the highest quality resolution available so that you can instantly go from low resolution to high resolution without deleting or hiding media.

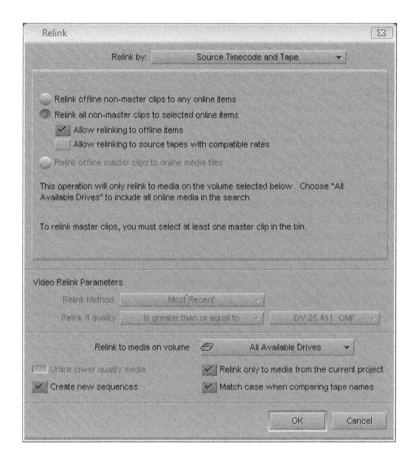

If you have media files offline that you know are on a connected drive you can simply Refresh Media Directories under the File menu. If this doesn't work then you will need to try relink. You can relink media when you have a master clip, subclip, sequence, or graphic that is offline. Highlight the object in the bin and choose "Relink" from the Clip menu. Your system searches the active MediaFiles folder for media that use the same tape name and timecode.

Rule 2—Tape Name and Project Name

Just because the tape name looks the same doesn't mean it is the same tape. There can be only one tape logged per project with a particular name. Every time you add a tape as "New" in the Tape Name dialog, you are creating a unique tape that is associated with that project. You can use "Tape 001" two different times in the same project, but the files are logged in separate projects and are always considered different tapes.

The project name is displayed as a column next to the tape name in the Tape Name dialog window. You can choose to show other project's tapes if you think the tape you need is in another project. To reduce confusion, all tapes for a project should be logged in the same project or in another project with the exact same name.

Having said all that, there is a way to bypass the project name. In the Relink dialog, uncheck the "Relink only to media from the current project." Then only the tape name is important and not the project name. You can choose whether the media files are from the same project and you can make sure that you will match the case of the tape name. The main use for unchecking the case of the tape name is when you are trying to link to an EDL (edit decision list) that has been imported. You can link to these clips if the tape name is the same, but most EDL formats change the original Avid name for the tape to something with all capital letters. An EDL may also truncate a long tape name, which is a good reason to use short or numerical tape names, as recommended earlier. If these two choices are checked, then the first time you try to relink, nothing may happen; however, if you are relinking to a terabyte or two of media, you will appreciate these choices when working with the tape names of dozens of projects online. Uncheck the choices and try the relink again.

Rule 3—Size Does Matter

A master clip cannot link to captured media files that are more than a few frames different from the master clip's start and end times, even though it has a common timecode and tape name. For various reasons, this rule became looser in later versions of the software, but only by a few frames.

Rule 4—Subclips Are Less Choosy

A subclip will relink to media files that are longer than the subclip. This is true even if the subclip is exactly the same length as a master clip that will not relink. Subclips are programmed to link to more media than the subclip start and end times.

Rule 5—Sequences Are Really Collections of Subclips

A sequence may relink when the individual master clips that are contained within it will not. Think of a sequence as many subclips.

Rule 6—Multipart Files Make Things More Complicated

A sequence or a subclip will not relink to a media file that is shorter than the media it needs unless it is a multipart file. If you are working with OMF media, a single capture can be constructed of several video media files (due to the 2-GB limit for OMF files). This can make relinking a bit more complicated, as it is possible that only part of the master clip will relink if one or more of those file parts are missing. If only one of several media files links to the sequence or subclip, then the clip will be partially online with the Media Offline slide displaying wherever a piece of media is missing.

Rule 7—Relinking Master Clips Is Different Than Relinking Sequences

In the Relink dialog, the system will gray inappropriate choices. If a master clip and a sequence are both selected, then you must choose which one you really want to relink.

Rule 8—Relinking a Sequence to Selected Master Clips Works Only in the Same Bin

In the Relink dialog box, the "Relink all nonmaster clips to selected online items" button can be used only when you have highlighted specific online master clips that you want to relink to a sequence. In older versions, this was called "Relink to Selected." There is no way to unlink a sequence, only forcing it to relink to other media, taking the original media offline or using the decompose function.

This Relink to selected online items option works only within a single bin. It is used primarily to force a sequence to relink to media at another resolution or from another project. To relink to clips from the Media Tool, the clips must first be dragged from the Media Tool into the same bin with the sequence. Everything in the bin to be relinked must be highlighted. Do not check this function unless you are specifically linking To Selected, or nothing will happen.

Unlinking

Unlinking is one of those powerful, dangerous, useful, and poorly understood functions that people know they should use but don't really know when. Sometimes a link must be deliberately broken using Unlink. In the Clip menu, highlight the desired clips, hold down Shift-Ctrl on Macintosh or Shift-Alt on Windows, and Relink becomes Unlink. Any media that have been captured for this master clip now become offline. The system considers this master clip as never been captured and is not linked to any media or to any sequences. The sequence linking is important because otherwise every use of this clip in any sequence will be changed. You want to change this one master clip, but you probably don't want every use of it to change, too. That is why Unlink is required as a safeguard.

You then can modify the duration of the clip, but you must recapture all the unlinked clips. Do not unlink media that have no timecode, and do not modify the master clip because you will be unable to batch recapture.

Unlink is extremely useful for multicam projects. After batch capturing Reel 1 from Camera 1, you can duplicate all the master clips, unlink them, and change the tape name (Reel 2, Reel 3, etc.). Now you can batch capture all the other camera angles. Just be sure to duplicate the original clips using Ctrl/Command+D and not Alt/Option + drag to another bin. You must be working with a true duplicate and not a clone of the master clip before you unlink.

There is no way to unlink a sequence using the Unlink command. Sequences have a loose link to media that allows them to change resolutions easily. The best way to unlink a sequence is to duplicate and decompose. You can throw away all the new decomposed master clips and just use the sequence. Because a sequence is loose about linking, you don't really need to unlink most of the time. You can just force the sequence to link to new material (Relink all nonmaster clips to selected online items with both media and sequence in the same bin), and it will automatically break the links to the old media.

Backing Up and Archiving

So, what is your most important job as an editor? Making a great edit, right? Well, if you are a freelancer without a staff IT archivist at your disposal your most important job is backing up! Why? Well, no matter how good an edit is, if the sequence is accidentally deleted or the system crashes, that edit doesn't exist and is useless to the client, no matter how brilliant it was. A good friend of mine commented once that he was the fastest editor in the world the second time he edited the job. In other words, when you lose your sequence you better be both brilliant and fast because you're about to do it all over again!

Let's look at some back-up strategies.

Daily Project Backups

In many respects, the most important thing you can buy for an edit bay is a spindle of CD-Roms. I feel strongly that you should back up your project at the end of every single day. Both Macintosh and Windows systems have CD-burning capabilities built in. Use them. At the end of each shift, grab a CD off the spindle, put it in the computer's optical drive, and copy your entire project to the CD. Then do it again. Label both carefully indicating the project, job/client code if one exists, and the date.

I recommend keeping two backups—one onsite and one offsite. One stays in the edit bay or at the facility. The other goes home with you every night. Not only is it a good backup practice—one recommended by virtually everyone—but you never know when your personal backup will save your client. Though this has thankfully never happened to me, I've spoken with many different editors who arrived one morning at the postfacility they were cutting at to discover that a burglary had occurred and the computer in their edit bay was stolen. One even arrived one morning to discover padlocks on the door and a foreclosure notice taped to the window. As the tapes are often the property of the client you can usually get those back without spending too much time in court. Good luck, however, convincing the judge to let you copy their project off the seized computer!

If you wish, you could also use a flash drive to make your personal copy. Remember, you are doing your client a service by backing up their work. You want them as a long-time customer of yours, right?

Long-Term Archival

When the project is over you'll want to back up the project. The easiest and most inexpensive way to back up media files is to *not* back them up. Instead, vault your tapes, any graphics or animations created, and a copy of the project. Remember that if you captured with timecode you can always recapture that material.

But with the proliferation of inexpensive terabyte and larger portable hard drives, you may well want to save some time and back up everything to a hard drive you can easily put on a shelf for long-term storage. Once again, if you want to quickly back up everything in a project use MediaMover to move all the project media into their own folders. Believe me, if you are in this business to make money and you want to back up your media you need MediaMover. Just go buy it already.

If you are backing up the media, be sure you also back up the project and any graphics or animations created. Copy these to the hard drive with the media and to a CD-Rom (or even a flash drive), and vault it as well.

Backing Up User Settings

You should be able to reproduce user settings in about five minutes, but some people resist it like a trip to the dentist. There is no way to lock your user settings to keep unauthorized folks from making a few "improvements" or just accidentally changing them, so it is a good idea to back them up somewhere safe and hidden. Many freelancers carry their user settings on a USB memory stick attached to their keychain.

When software versions change, especially substantial version changes, you should *always* remake your user settings. I know this sounds tedious, but recreating them solves many unusual and unpredictable problems, especially if your customized keyboard is complex. You may be trying to access menus that have been moved or deleted! Or there may be more subtle problems that don't immediately appear to relate to user settings. One of the first things Avid customer support asks you to do if you are getting unusual behavior, like a common feature suddenly not working, is to create a new project or user settings. If the software version has not changed recently, having a backup of your user settings may be enough to fix the problem, rather than having to recreate them from scratch.

Another reason to create new user settings whenever there is a version upgrade is that there are often new capabilities that you won't get *unless* you create a new user. For example, in version 3.0 there are new interface configurations and some very useful new default bin views. If you don't create a new user you won't get them.

Use Common Sense

Even though there is a lot to be in charge of when administering an Avid system, much of it is common sense and taking advantage of existing computer peripherals and software that make your job easier. Make sure the policy you decide on is followed uniformly. Make sure that all members of your staff are educated on the correct procedures as well as just a little troubleshooting. Then they can deal with those questions themselves during the night shift. You may want to consider creating a media management policy in writing and making sure your clients know it, even by getting them to sign it when beginning a project. If set up right, you will significantly reduce downtime and make it easier to diagnose and solve technical problems and missing media.

5

STANDARD-DEFINITION VIDEO FUNDAMENTALS

"The devil is in the details."

—German proverb

Back in the early days before nonlinear editing became the way of working, editors learned the craft a certain way—and learned about far more than why and where to make a cut. Before they could work in the edit bay they had to cut their teeth in the video dupe department, then work as an assistant, and so on. Of course, this meant they spent years—for very little pay—doing something other than what they wanted to do; but it was about more than simply "paying the dues," it was all about learning. And one thing video editors had to learn before they were allowed in the edit bay was *signal*.

Editors had to become intimately familiar with video signal, with a deep understanding of how to read, manipulate, and, most importantly, calibrate video voltages. Indeed, their first task as an online editor each shift was to "time" the room, making sure that the voltages throughout the room were calibrated. Because if they didn't, their show would likely look pretty terrible and be filled with all sorts of nastiness like horizontal shifts, vertical rolls, visual distortions at edits, luminance (brightness) and color shifts, and so on. The analog world was *hard*. Virtually anything could go wrong, and editors could be guaranteed that plenty would go wrong unless they became masters of the video signal.

The digital domain simplified things dramatically, but unfortunately we're now faced with an editing world where the details that are easy to ignore will truly bite you—and your clients—at the end. In this chapter we're going to dive deep into the world of video signal, starting first with the foundations of video and progressing into the digital world. Along the way I'm going to get incredibly technical. Indeed, some folks argue that video engineering documents are best suited for insomniacs. Though I don't agree with that statement, I certainly agree that this is the geekiest part of this book by a mile.

My goal with this chapter, and Chapter 6 on high-definition (HD) video, is not to turn you into an engineer, but to try to express, using as nontechnical a language as possible, the world of video signal. Trust me, if you want to go far in this business you need to know this stuff.

Signal Fundamentals

Despite the growing prevalence of digital video formats for production (including DV, Digital Betacam, XDCAM, and so on) it is important to remember that video began as an analog, voltage-based feat of engineering. Indeed, all digital video formats have analog video as their foundation. In addition, most of the video signals in today's broadcast or cable distribution facilities, with the exception of HD, spend some or all of their time in analog form, and must adhere to analog standards. So, even if you work exclusively in digital, it is crucial to learn the fundamentals of an analog signal.

To better understand video signal fundamentals, let's begin by breaking down the signal into its most basic components, beginning with a single-line black-and-white signal.

When an analog video camera captures an image, the image is measured, or sampled, as a series of voltages that describe the relative brightness of the image. The higher the voltage, the brighter that portion of the picture. Specific voltages are assigned to both black and white (these are often referred to as *video black* and *video white*). The entire image is measured by scanning, or sampling, the image from left to right, one line at a time from top to bottom.

Video Line Structure

The majority of the video signal is dedicated to the display of the picture and is referred to as the *active region* or *active picture*. Outside of the active region, a line of video contains additional information used to help synchronize and align the line so it is properly displayed. This region serves two purposes: to define the beginning and end of a line, and to turn off, or *blank*, the display so the electron gun can quickly *fly back* from the right edge to the left so that it can scan or display the next line. This synchronization area is referred to as the *horizontal blanking region*.

Two specific voltages are used in the horizontal blanking region:

- *Synchronization:* This is used briefly in a sync pulse to align the timing of the video line. The sync pulse contains a unique voltage that is significantly lower in voltage than any other portion of the video signal.
- *Blanking:* This is used throughout the blanking region (with the exception of the sync pulse). The voltage is the same as is used for video black.

To better understand what a video signal looks like, standardized test patterns are used. For the first part of our exploration into signal we will use a grayscale ramp test pattern. (A grayscale ramp test pattern is a horizontal gradient between black and white.) When displayed on a video monitor, the signal looks like the first figure shown here. The second figure shows the basic signal structure for a line of video displaying this test pattern.

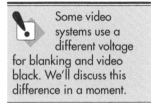

Some video systems use a different voltage for blanking and video black. We'll discuss this difference in a moment.

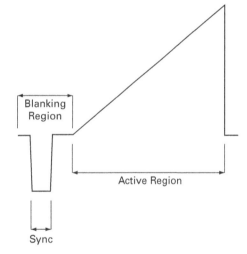

Video Line Voltages

Now let's add some actual voltages into the mix:

- In both NTSC and PAL, blanking is assigned a value of 0 volts.
- The blackest portion (video black) of the image is also assigned a value of 0 volts in PAL; NTSC video black will be described later.
- The whitest portion (video white) of the image is given a value of 714 millivolts (mV) in NTSC and 700 mV in PAL.
- The sync pulse is assigned a value of –286 mV in NTSC and –300 mV in PAL.

- The entire signal, from sync to peak white, has a range of 1 volt. This is often expressed as 1V P–P (one volt, peak to peak).

Though voltages are used as the unit of measurement in PAL, in NTSC the values of 714 mV and –286 mV don't lend themselves well to describing and measuring a signal. Therefore, the *IRE unit* was established to describe an NTSC signal. An IRE unit is equivalent to 1 percent of the range from blanking to peak white, or 7.14 mV. When expressed in IRE units, blanking is assigned a value of 0 IRE, peak white a value of 100 IRE, and sync a value of −40 IRE. The following image shows the signal structure for a line of video, this time with voltages and IRE units assigned.

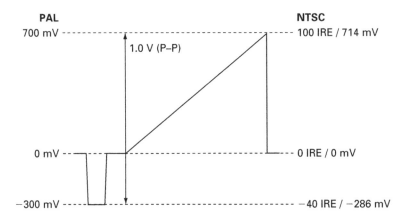

Black-Level Setup (NTSC Only)

In NTSC, a further signal distinction exists, that of black-level setup. Video black is raised slightly above the level of blanking.

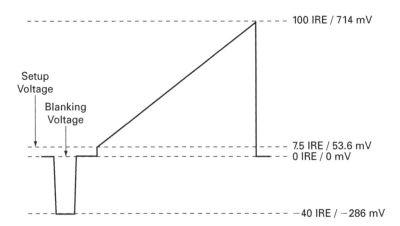

This was done to give television sets a distinct blanking signal apart from black. Black level is assigned a value of 7.5 IRE (53.6 mV). This means that blanking is actually *blacker than black*. Not all NTSC countries (e.g., Japan) use black-level setup. In those cases, black is assigned the same value as blanking (0 IRE) just as it is in PAL.

Black-level setup is only used in NTSC analog composite and component formats. It is not used in any digital component format.

Adding Color

Black-and-white images are easy to record as a video signal as each level of brightness is simply assigned a voltage. Color complicates matters somewhat. Video cameras capture color information by breaking down an image into three primary colors: red, green, and blue (RGB). (These three colors are known as the additive primaries and are the primary components of light.) By describing the percentage of each of these colors, we can record and reproduce a large portion of nature's colors.

When a camera captures a color scene, the color information is not captured linearly. This nonlinearity is due to the fact that virtually all video cameras are less sensitive to changes in darker areas than they are to changes in lighter areas. This nonlinearity is referred to as a camera's *gamma response curve* or simply as its *gamma*. When recording a video signal we correct for this nonlinearity, or gamma, and use the prime mark (') to indicate that the signal has been *gamma-corrected*. Therefore, in video we refer to these signals as the R', G', and B' signals.

To best understand how a color signal is created we will use a standard set of color bars. The next figure shows color bars with 100 percent (of peak) chroma saturation on all three color channels. The following illustration shows the signal produced from this color bars pattern for red (R'), green (G'), and blue (B').

Notice that each of these signals is full bandwidth. Unfortunately, recording a color image as R'G'B' means that we need to store three times the amount of information than we do with a black-and-white image. This presented the engineers who developed the color system with a big problem: There simply wasn't enough bandwidth in the video signal to record that much information. Additionally, the engineers wanted to create a color signal that black-and-white televisions could interpret and display.

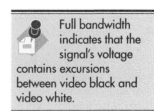

Full bandwidth indicates that the signal's voltage contains excursions between video black and video white.

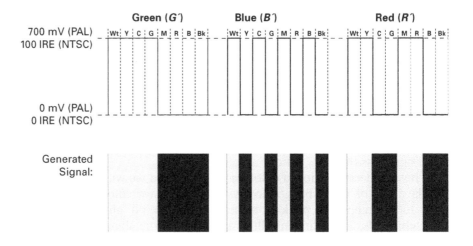

This method of storing color information was based on standards defined by the CIE (Commision Internationale de L'Éclairage) in 1931.

The Component Video Solution

The solution was to store the signal in some other way. The signal has to start as an R′G′B′ signal and end as an RGB signal (in order for the monitor to display it). The challenge was to develop a method that took less bandwidth but would be easy to encode and decode.

The engineers decided to store a full-bandwidth luma (black-and-white) signal and two color-difference signals of lower bandwidth. The human eye is more sensitive to differences in brightness than it is to differences in shades of color. By dedicating the majority of the signal to luma information, the engineers were able to take advantage of the way our eyes work. The color information must be stored in two additional *color-difference* signals, making for a total of three signals. Three signals are required to ensure that the original R′G′B′ information can be converted back to R′G′B′ for display on the monitor.

Storing a video image in this color-difference form has two significant advantages over storing it as R′G′B′:

- Substantially less bandwidth is required as only one high-bandwidth signal is required for the luminance as opposed to three high-bandwidth signals for R′G′B′.
- Gain distortions in any one of the component signals have a less-detrimental affect on the picture. A low level on one channel in a color-difference signal will only produce subtle changes in brightness, hue, or saturation. A gain distortion in R′G′B′ will produce significant color shifts throughout the entire image and can even produce "illegal" colors that exceed what is allowed for broadcast.

The Luma Signal

Luma, often referred to as Y′, is created by combining the red, green, and blue signals. They aren't combined in equal parts. This is because the human eye is more sensitive to some colors than others. We see the greatest amount of detail in greens, less detail in reds, and very little detail in blues. Therefore, the luma signal is primarily composed of 58.7 percent G′, 29.9 percent R′, and 11.4 percent B′. This can be expressed as:

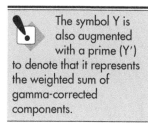

The symbol Y is also augmented with a prime (Y′) to denote that it represents the weighted sum of gamma-corrected components.

$$Y' = 0.587\ G' + 0.299\ R' + 0.114\ B'$$

The following illustration shows the luma portion of the signal for the 100-percent color bars pattern. *Note:* The steps in the "staircase" are not equal. This is because of the percentage of red, green, and blue used. For example, the first bar, white, is composed of all three values $(0.587\,G' + 0.299\ R' + 0.114\ B')$ and the second bar, yellow, is only composed of green and red $(0.587\,G' + 0.299\,R')$.

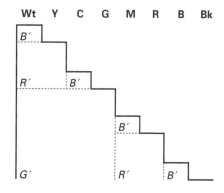

The Color-Difference Signals

In addition to luma, two color-difference signals are required. The color-difference signals store the color information that is different from the luma information. These signals are created by subtracting luma from one of the original red, green, or blue signals. This way, we can recreate the original signal by adding together the color-difference signal and the luma signal.

As the luma signal is primarily made up of green, subtracting luma from green doesn't yield a very useful signal. Therefore, one color-difference signal is created by taking blue and subtracting luma from it (B′–Y′); the other by taking red and subtracting luma from it (R′–Y′).

Using the same 100-percent color bars pattern, the following figure shows the created B′–Y′ and R′–Y′ signals. These values are then normalized so that the peaks for B′–Y′ and R′–Y′ are identical. NTSC and PAL use different voltage normalizations, as shown in the illustration.

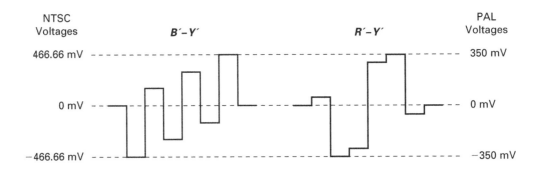

In summary, the three signals used for a color video image are Y′, B′–Y′, and R′–Y′. From these signals, RGB can be recreated mathematically by circuits in the monitor.

Every method of storing and transmitting a color video signal (composite, S-video, component, and digital) uses the previously discussed method to create the key parts of the video signal. Now, let's examine each standard to see how they differ.

Composite Video

Composite video is unique among all the analog video formats in that it is the only one actually broadcast over the air. All of the others are only transmitted between equipment within a facility.

Composite video is the oldest method of storing a color video signal. It also has the least available bandwidth and the lowest quality. To reduce the amount of space required for our two color-difference signals, we take advantage of another characteristic of the human eye: We cannot see fine detail in color changes as easily as we can see fine detail in brightness changes. Therefore, we can reduce bandwidth of the two color-difference signals without noticeably degrading the image as long as we don't reduce the bandwidth of the luma signal. Additionally, composite color requires that the three signals—Y', B'–Y', and R'–Y'—be combined into a single signal. To do this, a method known as *encoded color* was developed.

Encoded Color

Encoded color uses a well-established concept: amplitude modulation (which is used for AM radio). Amplitude modulation uses a high-frequency carrier wave whose amplitude (height) is varied by the addition of another frequency. A carrier wave is simply a high-frequency sine wave of a specific frequency. When an additional signal is applied, the height of the sine wave varies with the voltage of the added signal—the greater the voltage, the greater the amplitude of the sine wave at that point in time. Additionally, the carrier wave frequency must be high enough to capture all of the color information.

Amplitude modulation is a form of analog sampling. Each period of the carrier wave carries one voltage, or sample. A period of a sine wave starts with no voltage at 0°, travels to peak positive voltage at 90°, returns to no voltage at 180°, travels to peak negative voltage at 270°, and returns to no voltage at 360°.

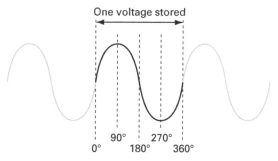

Due to the different frame rates and line counts of the NTSC and PAL standards, the two formats use different carrier wave frequencies for their encoded color:

- NTSC uses a carrier wave frequency of 3.58 MHz.
- PAL uses a carrier wave frequency of 4.48 MHz.

Because there are two signals (B'–Y' and R'–Y') to be encoded, one is carried on one carrier wave and the other on a second carrier wave that is offset 90° from the first. (This is technically known as *phase quadrature* and the entire process as *quadrature amplitude modulation*, or QAM.) This encoded signal is commonly referred to as chroma.

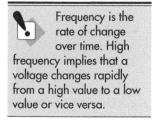

Frequency is the rate of change over time. High frequency implies that a voltage changes rapidly from a high value to a low value or vice versa.

In addition, the B'–Y' and R'–Y' signals are scaled and bandwidth-limited. Once scaled and limited these signals are often referred to as U and V, respectively.

Combining Luma and Chroma

We now have only two signals, luma and chroma, which have to be combined. The luma and chroma signals from our 100-percent color bars are shown in the next figure for reference.

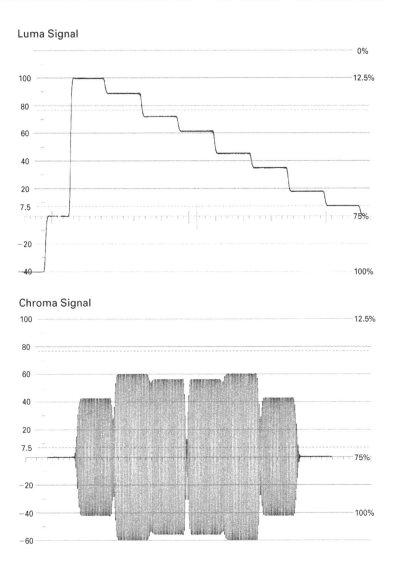

Luma Signal

Chroma Signal

If we compare the luma and chroma signals, you will observe that the luma signal contains primarily low-frequency (the rate of change over time) information while the chroma contains primarily high-frequency information. The only high-frequency information in the color bars test pattern is at the transition between one bar and the next. Though this transition is quite rapid, it is not immediately repeated.

Indeed, it can be stated that luminance signals are primarily composed of low-frequency information. This is true not just for test patterns, but for almost all video signals. (Luminance signals can contain some high-frequency information. This is most typically seen in high-contrast patterns such as a plaid jacket. Accurate representation of these signals is unfortunately not always possible in composite video systems.)

Since the luma signal contains few high-frequency data and the chroma signal is entirely composed of high-frequency data, the two signals can be combined together (the chroma signal is summed with the luma signal) with little risk of signal contamination. (Since the chroma is carried in a carrier wave and is the secondary of the two signals, it is referred to as a *subcarrier*.)

Additionally, an unmodulated portion of the carrier wave is placed on the back porch (see sidebar) of the blanking interval. This portion, known as the *color burst*, is used when decoding the composite video signal to isolate the encoded color from the rest of the signal.

Bar Bets and Video Signal

When you are talking with video engineers you'll often hear them refer to the right section of blanking after sync as the *back porch*. Likewise, they'll refer to the left section of blanking prior to sync as the *front porch*. Where did these terms come from? Believe it or not, from the design of the typical house in a hot and humid climate (such as the southeast United States). These houses always have a front and back porch (for living on during the cooler evening hours) and a single hallway flowing down the middle of the house (known as the breezeway). This house design utilized a concept known as negative air pressure to draw air through one of the two openings, through the house, and back out the other opening. This negative airflow would pull a significant amount of air through the house and dramatically cool what would otherwise be an unlivable environment.

As crazy as it sounds, this is literally where these terms came from. Some engineers, especially down in the South, still refer to the sync pulse as the breezeway. So when you're out late with an especially geeky set of video editors, feel free to pull this arcana out and use it to get someone else to buy the next round!

When we combine the luma and chroma signals from the 100-percent color bars, the signal in the following figure is produced.

Notice that the chroma in the summed composite signal has been raised by the luminance voltage. The composite amplitude of the signal is the sum of the luma and chroma voltage.

Peak Composite Amplitude

The 100-percent color bars pattern produces a peak composite voltage of:

- 131 IRE (NTSC) or 1000 mV (PAL).

The 100-percent color bars pattern produces a minimum composite voltage of:

- −23 IRE (NTSC) or −300 mV (PAL).

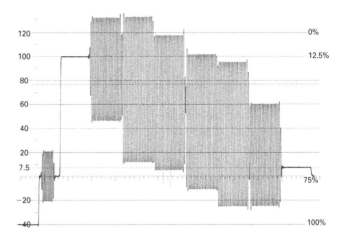

Although this is the maximum allowable voltage according to the composite video specification, the 100-percent color bars pattern's peak voltage exceeds what most broadcasters will accept for a composite format signal.

For this reason, this 100-percent color bars pattern is usually replaced by a 75-percent color bars pattern where the maximum chroma is only 75 percent of the allowable peak. The 75-percent color bars pattern creates a peak composite amplitude of:

- 100 IRE (NTSC) or 700 mV (PAL).

This composite signal has the same peak amplitude as the luma signal. The following illustration shows the signal produced by a 75-percent color bars test pattern. *Note:* We've configured the scope to display both the full composite signal and the luma signal side by side. Notice that the luma voltages in the 75-percent pattern are different than the 100-percent color bars pattern.

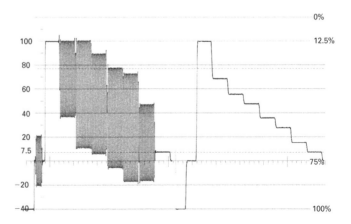

The entire system is quite ingenious and can be encoded and decoded entirely using analog circuitry.

Composite Video Limitations

Composite signals are not without their problems. As the color-difference signals are kept separate by a difference in phase, very small phase distortions can cause large color distortions in a picture. These phase distortions are so common in NTSC signals that a subcarrier phase adjustment (known as *hue*) is standard on every television set.

Many other problems exist with composite signals and are principally a result of the encoding, overlaying, and decoding of the chroma signal. This is often referred to as the composite footprint and manifests itself in chroma crawl, chroma edge inaccuracies, cross-color moiré patterns, and so on. For example, the plaid jacket mentioned earlier is represented by high-frequency luma information. When the composite signal is decoded, this high-frequency luma data may be converted to chroma information. This results in cross-color moiré patterns. If the signal is reencoded to composite, the jacket pattern is left as chroma and some luma detail is lost. On the next decode, some of the remaining luma information is converted to chroma and even more luma data are permanently lost.

Composite Video Differences in PAL

As just mentioned, one of the key limitations of the composite video format is that the subcarrier phase cannot be readily deduced, causing hue shifts in the decoded image. The designers of the PAL video format solved this problem by inverting the subcarrier phase with every other line of video.

Decoding circuitry was then designed to use these phase inversions to calculate the correct subcarrier phase, eliminating the need for a subcarrier phase (or hue) adjustment by the end user.

In addition to PAL, the SECAM format is also used in some European countries for transmission. Studio production in these countries, however, is often done in PAL or 625-line component as SECAM is a transmission-only format.

Video Frame Rates

Prior to the introduction of color, American television operated at a frame rate of 30 frames per second (fps) and European television operated at a frame rate of 25 fps. These two frame rates were not chosen arbitrarily, but instead were chosen to correspond to the alternating current (AC) power frequency used in the host countries (60 Hz and 50 Hz, respectively).

The developers of the NTSC format discovered that when color was added to the signal, the difference between the audio and color subcarrier frequencies introduced a noticeable dot pattern across the image. To reduce and hopefully eliminate this dot pattern, they slowed the frame rate slightly by a factor of 1.001, resulting in a new frame rate of 29.97 fps. Fortunately, this adjustment was compatible with existing black-and-white televisions and was adopted. Unfortunately, this change has led to a wide variety of problems and compromises in video production, post-production, and transmission that continue to this day, even into the HD formats.

As the PAL video format did not have to be compatible with earlier European black-and-white transmission formats, the developers of the PAL format were able to avoid this problem and design a system that would support the 25-fps rate.

S-Video

S-video, also referred to as Y/C, is a simple variant of composite video. The two color-difference signals are encoded using QAM, just as with composite video. The only real difference is that the luma and chroma are not multiplexed (combined), but are left as two separate signals. This method eliminates the problems inherent in combining and separating luma and chroma, but it still has a reduced chroma bandwidth.

Y/C signals are extremely advantageous in composite video situations as they allow the encoded color to be transmitted from one piece of equipment to another without being multiplexed and demultiplexed over and over again. This is critical because the composite footprint becomes more and more apparent each time the signal is demultiplexed and remultiplexed.

Though Y/C signals have been used for years, they became the foundation of the consumer and industrial format S-VHS. Matsushita, the developer of S-VHS, coined the phrase S-video when they introduced the format, and the name continues to be used.

Component Video

Earlier we said that all video starts as RGB color and is translated into luma (Y′) and two color-difference signals (B′–Y′ and R′–Y′). The component video format keeps the three signals (Y′, B′–Y′, and R′–Y′) separate and does not encode and overlay the color information over the luma. The two signals are scaled just as they are in a composite signal, but the scaling applied is different

Luma Signal

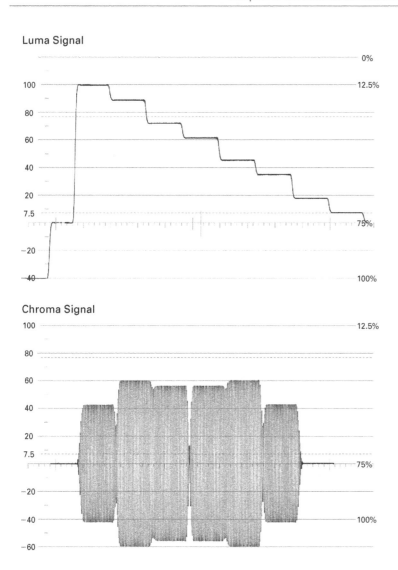

Chroma Signal

than that used for composite signals. The color-difference signals are referred to as P_B and P_R in analog component systems.

The following diagram is the signal resulting from a set of full-field 100-percent (of peak chroma) color bars for the 525-line and 625-line formats. Note that the voltages for Y', P_B, and P_R are different for the two formats. The 625-line format uses the EBU N10 standard. Though the 525-line format can also use EBU N10 (which is also referred to as SMPTE N10), the Betacam component standard is more common. The Betacam component maintains the NTSC composite voltages for luma to make it easier to transcode component video to composite and vice versa. All Avid systems use the Betacam component standard.

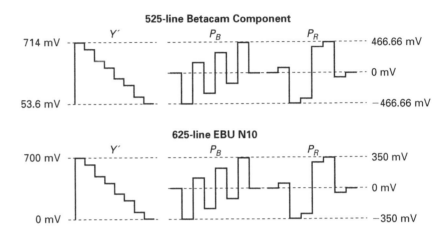

NTSC and PAL in the Component World

Technically, the terms NTSC and PAL are only valid if used to describe a composite video format. Component video formats instead are named by the number of scan lines in the format. Therefore, the proper terms are *525-line* (instead of NTSC) and *625-line* (instead of PAL). We will discuss scan line counts in greater detail in the next section.

Also notice that a 525-line component signal is measured in millivolts, not IRE. IRE measurements are only valid for composite signals.

By keeping the three signals separate we eliminate many of the problems inherent in composite video. For example, pulling a clean chroma key from a composite signal is extremely difficult, particularly around the edges of an object. This is due to the inevitable footprint of composite encoding. However, a clean chroma key is fairly easy to pull from a component signal.

Additionally, component video does not have to undergo the amount of signal compression needed for composite video and it can handle a much greater bandwidth of color information. Component video does, however, have its challenges. Any time you have three signals traveling through three distinct wires you run the risk of variances in both gain and timing for the three signals. Therefore, it is critical that component analog equipment is connected using three cables of identical length.

Video Frame Structure

Up until now we've focused primarily on the picture portion of the video line. Now let's take a look at the rest of the video signal.

Blanking Interval

As we alluded to at the beginning of the module, a portion of the line is reserved for synchronization information called the *blanking interval*. The following illustration displays this interval in detail.

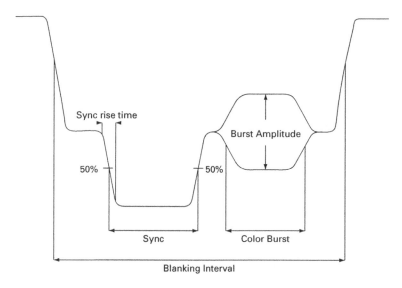

The blanking interval contains several critical synchronization components, each of which must have specific timings and/or amplitudes. Table 5.1 lists the critical components and their respective timings and/or amplitudes in both NTSC and PAL. The information for NTSC is derived from the SMPTE 170M specification and the information for PAL is derived from the ITU System B specification. These two specifications codify the NTSC and PAL analog video formats.

The blanking interval derives its name from its primary purpose: to shut off, or blank, the video signal between each line of video. This is required because the video frame is displayed one line at a time from left to right. The blanking interval allows the signal to rapidly *fly back* to the left edge of the screen at the end of each *active line period*.

Table 5.1 Blanking Interval Timings and Amplitudes

Blanking Component	NTSC (SMPTE 170M)	PAL (ITU System B)
Blanking interval duration	10.9 μs ± 0.2 μs	12.0 μs ± 0.3 μs
Sync rise time	140 ns	140 ns
Sync duration	4.7 μs ± 0.1 μs	4.7 μs ± 0.2 μs
Sync amplitude	−40 IRE (286 mV)	−300 mV
Color-burst duration	9 subcarrier cycles ± 1	10 subcarrier cycles ± 1

μs = microsecond, or one-millionth of a second; ns = nanosecond, or one-billionth of a second; subcarrier cycle = the time it takes the subcarrier to complete one period.

Scan Line Structure

The total number of lines in the video frame is different for NTSC and PAL. NTSC uses 525 lines for each frame and PAL uses 625 lines. Though the majority of these lines are used for picture (and are referred to as the active picture area), some lines are reserved for vertical synchronization and are referred to as the vertical blanking interval. (We'll discuss these lines in greater detail in a moment.)

In addition, due to limitations of technology and bandwidth (the amount of space available to store the video signal), only half of the frame is scanned at a time. The first pass through the image scans every other line. The second pass scans the skipped lines. These two scans combine to create the whole image. This scanning sequence is known as *interlaced* scanning. Each scanning pass is called a *video field*. The following illustration shows the scanning pattern for the active picture area for both NTSC and PAL.

Both NTSC and PAL begin and end the active picture area with a half line. The other half of the line is part of the vertical blanking interval. Also note that the top line in NTSC is from field 2 while the top line in PAL is from field 1.

NTSC Interlacing

PAL Interlacing

The Boy Who Invented Television

At the beginning of the twentieth century, inventors and companies around the world were trying to figure out how to transmit images electronically over the air. The irony was that a 14-year-old boy had already figured it out. Philo T. Farnsworth was harvesting potatoes on his family's farm, endlessly driving the harvester row by row back and forth through the field, when inspiration struck. What if one were to "draw" a picture on a television screen line by line just as one would plow a field? Just as the eye could resolve the field from the plow lines, he imagined that the eye could stitch the transmitted lines back together and see the picture. He further imagined what he called his "image dissector," which used an electron gun to scan an image and then redraw it in another location. (Other inventors used large spinning mechanical disks to scan the image.)

The year was 1922, a full year before anyone else had come to a similar conclusion. The sad reality was that it wasn't until July 1957 that the television industry finally publically acknowledged his contributions—and that acknowledgment came on the game show "I've Got a Secret." Between his invention and that game show were decades of legal battles between Farnsworth and the massive RCA corporation who wanted sole ownership of the invention. The tale is a fascinating one, and if you want to read more about it, visit www.farnovision.com.

Vertical Blanking Interval

Let's look at the vertical blanking interval (VBI) in greater detail. The interval is composed of two sections: a reserved area with specific vertical synchronizing and equalizing pulses, and a section available for special vertical interval signals. These signals can include such information as vertical interval timecode

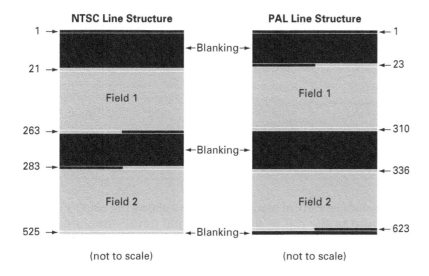

(VITC), closed-captioning data, and other types of information as defined by the broadcaster. The following illustration and Table 5.2 show the line structure for both NTSC and PAL. To enhance readability, the blanking sections have been written in italics.

Table 5.2 NTSC and PAL Line Structure

NTSC Line Structure		
Line/Field	**Left Half of Line**	**Right Half of Line**
1–9/f1	*Blanking: EQ and sync*	*Blanking: EQ and sync*
10–20/f1	*Blanking: vertical interval*	*Blanking: vertical interval*
21–262/f1	Active picture	Active picture
263/f1	Active picture	
263/f2		*Blanking: EQ and sync*
264–271/f2	*Blanking: EQ and sync*	*Blanking: EQ and sync*
272/f2	*Blanking: EQ and sync*	*Blanking: empty*
273–282/f2	*Blanking: vertical interval*	*Blanking: vertical interval*
283/f2	*Blanking: vertical interval*	Active picture
284–524/f2	Active picture	Active picture
525/f2	Active picture	Active picture
PAL Line Structure		
Line/Field	**Left Half of Line**	**Right Half of Line**
1–5/f1	*Blanking: EQ and sync*	*Blanking: EQ and sync*
6–22/f1	*Blanking: vertical interval*	*Blanking: vertical interval*
23/f1	*Blanking: empty*	Active picture
24–310/f1	Active picture	Active picture
311–312/f1	*Blanking: EQ and sync*	*Blanking: EQ and sync*
313–318/f2	*Blanking: EQ and sync*	*Blanking: EQ and sync*
319–335/f2	*Blanking: vertical interval*	*Blanking: vertical interval*
336–622/f2	Active picture	Active picture
623/f2	Active picture	
623/f1		*Blanking: EQ and sync*
624–625/f1	*Blanking: EQ and sync*	*Blanking: EQ and sync*

The longer vertical blanking time also allows the electron beam to return to the top of the screen. (Due to the structure of the beam, the vertical retrace is much slower than the horizontal retrace.)

Subcarrier Synchronization (Composite Video Only)

In addition to the syncronization signals in the vertical interval, the NTSC and PAL composite video formats contain a phase synchronization structure for the color subcarrier. This structure is four fields long in NTSC and eight fields long in PAL. This relationship is referred to as *SCH* or *subcarrier-to-horizontal* phase. Editing in a pure-composite environment required editors to maintain the SCH alignment of these four- or eight-field structures or color errors could occur. Fortunately, this is not necessary in component video formats as the chroma signal is not contained in a modulated subcarrier.

Introduction to Digital Video

As we learned earlier in this chapter, *sampling* is the fundamental process of creating a video signal by converting what the camera sees into a series of voltages. Though these analog voltages can accurately represent the image, there are many problems with analog recordings. Any analog signal is subject to voltage errors or loss that can be introduced while transmitting the signal from one location to another. In addition, the process of merely reading and rerecording the voltages can introduce small generational errors that over time can dramatically reduce the quality and accuracy of the signal. And finally, mixing multiple analog signals carried over different cables in a production environment can introduce timing errors that will further degrade the signal.

Digital signals, on the other hand, can be much more resistant to such errors and degradation. Digital data can be packaged and sent over almost any distance with no appreciable loss in quality.

When the engineers began to develop digital video, they built on what had already been established for analog video. Just as was the case with analog, storing the signal as R'G'B' would be an inefficient approach and they decided to use the same Y', B'−Y', and R'−Y' signals as they had for analog.

Standards exist for both digital composite and digital component video. However, the digital composite video format, implemented in the tape formats D2 and D3, has fallen by the wayside in favor of digital component video and will not be discussed.

Digital Component Video

Digital component video captures the three video components—Y', B'−Y', and R'−Y'—as three separate signals. The digital derivation of Y′ from R′G′B′ uses the same weighted function as used in analog video:

$$Y'_{601} = 0.587\ G' + 0.299\ R' + 0.114\ B'$$

The Y′ component is subscripted with 601 to indicate that the luma component is derived using the values ascribed in ITU-R BT.601, the international specification for standard-definition (SD) digital video. High-definition video uses a different derivation of luma.

The B′–Y′ and R′–Y′ signals are low-pass filtered, bandwidth reduced, and are referred to as C_B and C_R, respectively.

Sampling the Analog Video Signal

Since digital video was based on the analog video standards of the day, when converting an analog video signal to digital, two questions must be answered:

1. **What portion of the signal should be captured?** Though the obvious answer would be to capture the entire signal, that isn't always the best approach. Most of the blanking interval (both horizontal and vertical) contains synchronization signals that can be more efficiently represented in the digital domain using small data timing blocks. Therefore, the designers decided to sample the active picture and omit most of the blanking interval. We'll look in detail at the portions to be captured later in this section.

2. **At what level of detail should the material be captured?** This question concerns sampling resolution. Sampling is the process of capturing analog information for measurement and is similar to looking at an image through a wire mesh. Each open hole in the mesh is a single sample. All of the image detail within the hole is averaged to create a single color value for the sample. The finer the sampling, the more detail that can be measured and stored.

If the sampling rate is not high enough to capture the relevant information, errors can be introduced. These errors are known as *aliasing* and can result not only in a reduction of detail, but in worst-case situations, in wholly incorrect data being measured.

If you recall from our earlier discussion of encoded color, a sampling frequency of 3.58 MHz (NTSC) or 4.48 MHz (PAL) was used to sample the chroma information in the analog composite format. As the luminance signal contains significantly more information than the chroma, a much higher sampling rate must be used to ensure that the luminance data are accurately captured.

Digital Sampling Frequency

A joint SMPTE/EBU (Society of Motion Picture and Television Engineers/European Broadcasting Union) taskforce worked to define a common sampling method that would work for both 525-line (NTSC) and 625-line (PAL) systems. They settled on a luma sampling rate of 13.5 MHz. This sampling method was codified in ITU-R BT.601.

A sampling rate of 13.5 MHz results in a total of 858 samples for 525-line and 864 samples for 625-line across an entire line of video (including the blanking interval). Since the analog blanking interval is not really necessary to synchronize a digital signal, only 720 samples are actually used to capture the actual picture information. The sampled region is known as the *digital active line.* The following illustration shows this sampling of the video signal.

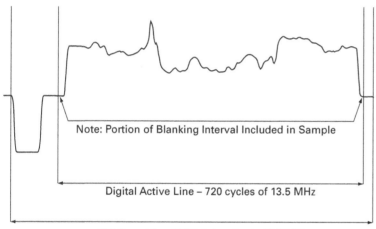

Note: Portion of Blanking Interval Included in Sample

Digital Active Line – 720 cycles of 13.5 MHz

525-line video: 858 total cycles in 13.5 MHz
625-line video: 864 total cycles in 13.5 MHz

Notice that the digital active line is slightly longer than the analog active region and that a small portion of the analog blanking interval is sampled. This is done deliberately as the transition (or *rise time*) in and out of blanking is considered a critical timing in broadcast environments. Whenever an analog signal is sampled, this blanking region is preserved.

Since the entire analog horizontal blanking interval is not sampled, some other mechanism must be used to synchronize the line data. Two special digital signals, *SAV* (start of active video) and *EAV* (end of active video) packets, are used. These are placed at the beginning and end of the active picture region in each sampled

This blanking region is only required for analog-originated signals. Preserving this region is optional for digital cameras and other digital-originated signals.

line. The remainder of the digital line (between the EAV and SAV packets) is used to store audio and other ancillary data. This region is often referred to as the HANC (*horizontal anc*illary).

Even though the horizontal blanking interval is not sampled, a standard analog sync reference signal is used when synchronizing digital equipment with other digital or analog equipment.

4:2:2 and Other Sampling Methods

As mentioned earlier, different sampling rates are used when sampling the luma and color-difference signals. In digital composite systems, the sampling rate of four times the frequency of the subcarrier, or $4f_{SC}$, was used to sample the composite signal. Out of this early system came the convention of using the numeral 4 to indicate that full-rate (13.5 MHz) sampling was used. It was then further decided that the numeral 2 would indicate half-rate (6.625 MHz) sampling.

4:2:2 Sampling

ITU-R BT.601 specifies that the luma (Y′) is sampled at the full-rate of 13.5 MHz and the chroma (C_B and C_R) is sampled at the half-rate of 6.75 MHz. Using the numbering convention just noted, this format is described as having 4:2:2 sampling. 4:2:2 sampling results in 720 Y′ samples and 360 C_B and C_R samples per digital active line. The following illustration shows what 4:2:2 sampling looks like in two dimensions.

Most component digital equipment uses 4:2:2 sampling. Two other sampling systems, 4:1:1 and 4:2:0, are also used in digital video, primarily in MPEG-based systems. Both use full-rate sampling for the luma. Their differences come in how the color-difference signals are sampled.

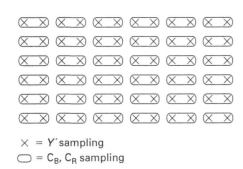

$\times$ = Y′ sampling
$\bigcirc$ = C_B, C_R sampling

$\times$ = Y′ sampling
$\bigcirc$ = C_B, C_R sampling

4:1:1 Sampling

In this sampling system, the color-difference signals are sampled at a quarter-rate, or 3.375 MHz, resulting in only 180 C_B and C_R samples per digital active line. 4:1:1 is primarily used in MPEG-based systems, including consumer DV, DVCAM, and DVD. The following illustration shows what 4:1:1 sampling looks like in two dimensions.

4:2:0 Sampling

In this sampling system, the color-difference signals are sampled at half-rate, but are only sampled on every other line. Though the exact placement of the C_R and C_B samples varies in different 4:2:0 implementations, the MPEG-2 and 625/50 DV formats use co-sited sampling: where Y′ is sampled on every line, C_B is only sampled on every odd line, and C_R is only sampled on every even line. 4:2:0 sampling is used in Panasonic DVCPRO 25 systems and some PAL DV systems. The following illustration shows what 4:2:0 sampling looks like in two dimensions.

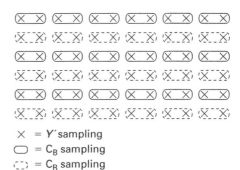

$\times$ = Y′ sampling
$\bigcirc$ = C_B sampling
$\vdots$ = C_R sampling

Representing Voltages Digitally

If you recall from the previous module, 525-line and 625-line systems use different voltage ranges for storing picture information. To simplify matters, digital component video uses a single-voltage range for both 525-line and 625-line systems. The voltage range used is referred to as SMPTE N10 and specifies that video black is assigned a voltage of 0 mV and video white a voltage of 700 mV. It further specifies that there is no setup in 525-line systems.

The analog voltage is digitally sampled and stored as either an 8-bit or a 10-bit number.

Luma (Y′) Sampling

The following illustration shows the digital values used to sample the luma signal. *Note:* Even though component digital only captures the video signal between the SAV and EAV packets, the blanking interval is included to provide an example of the sampled signal headroom and footroom.

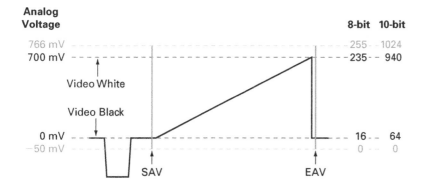

Notice that a reasonable amount of headroom and footroom is provided beyond video black and video white (66 mV of headroom and 50 mV of footroom, respectively). This headroom and footroom is shown in the previous illustration in gray. Due to the nature of digital video, filtering and compression can cause *blooming* and *ringing* at the signal's limits. Giving the signal generous headroom and footroom can help eliminate or dramatically reduce these problems. In addition, the footroom allows for the digital sampling of luma key elements, a critical requirement in broadcast environments at the time the standard was developed.

Don't Confuse Your Analog and Digital Voltages!

Remember that 525-line digital uses the voltage range of 0 mV–700 mV, while 525-line analog uses a voltage range of 53.6 mV–714 mV. Because of this, care must be taken when switching between the two formats and measuring signal information on an external scope. Assuming the analog and digital formats use the same voltages is a common mistake made when first working with digital video systems.

Color-Difference (C_B and C_R) Sampling

The color-difference signals are similarly sampled, with the values of −350 mV and 350 mV used for the nominal peak chroma levels. As with the luma signal, a reasonable amount of signal headroom and footroom are provided (approximately 50 mV for both headroom and footroom). The following illustration shows the digital values used to sample the color-difference signals. *Note:* Though the C_R signal is shown in the illustration, the same voltage range and bit values are used for both C_R and C_B.

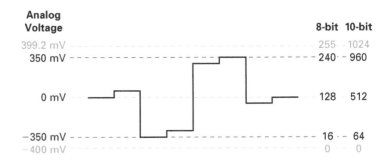

Reserved Samples

The lowest four digital values (0–3) and the highest four digital values (2528–2558 or 102,110–102,410) are reserved (in the Y', C_R, and C_B signals) for synchronization purposes.

Video Line Sampling

As mentioned earlier in this module, digital video systems do not need to sample the majority of the analog signal's blanking interval. Instead, this time interval is used for ancillary digital data. Only the active picture lines are sampled. That said, the actual lines sampled vary depending on the digital implementation. Though ITU-R BT.601 provides the specification for digital line sampling and encoding, it does not define which video lines will be sampled. As a result, various digital video formats (such as Digital Betacam, DV, etc.) have their own specification for the video lines that will be sampled and stored.

The video line sampling most commonly used is specified in ITU-R BT.656. This specification provides for 486 digital active lines in 525-line video and 576 digital active lines for 625-line video. Digital tape formats including Digital Betacam, D1, and D5 conform to this specification. Table 5.3 lists the digital line sampling specified in ITU-R BT.656 for the 525-line and 625-line formats.

Table 5.3 ITU-R BT.656: 525-Line and 625-Line Sampling

Sampling Region	525-Line	625-Line
Field 1: start of active picture	21	23
Field 1: end of active picture	263	310
Field 2: start of active picture	283	336
Field 2: end of active picture	525	623

Notice that only the active picture is sampled; the vertical blanking interval is not sampled from the analog. The serial digital interface standard specifies that this region can be used to transmit ancillary data such as synchronization, audio, or other information. This region is often referred to as the VANC (*vertical ancillary*). It is not required by the standard that this area be stored on tape or disk.

All two-field Avid media, with the exception of the DV resolutions, conform to the ITU-R BT.601 and ITU-R BT.656 specifications, providing for 4:2:2 sampling of a 720×486 (525-line) or 720×576 (625-line) frame.

Digital Frame Structure

Let's take a look at the structure of the digital frame. The following illustration shows the line structure of the interlaced frame for both the 525-line and 625-line formats.

Notice the field ordering of the 525-line and 625-line formats. The field ordering follows the same structure as the analog

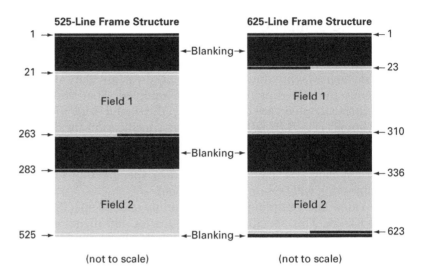

format. In the digital domain, field ordering is often referred to by the *topness* of the active frame. If the first spatial line of the frame is from field 1 the format is said to be field 1 ordered, and if the first spatial line of the frame is from field 2, the format is said to be field 2 ordered. Therefore, 525-line video is *field 2 ordered* and 625-line video is *field 1 ordered*.

Unfortunately, this terminology is not universal and other terms are used by various manufacturers and software programs. Some number the field by its *temporal* position in the frame data stream instead of by the topness. When the frame is viewed temporally from the first to the last line, field 1 is said to be in the *upper* position and field 2 in the *lower* position. The field ordering is still measured by the topness, but the upper and lower terms are used instead of field 1 and 2. Therefore, 525-line video is said to be *lower-field ordered* and 625-line video is said to be *upper-field ordered*.

Finally, some programs refer to the fields not as field 1 and field 2, but as *odd* and *even* fields. (Thankfully, nearly all graphics and compositing programs today start counting fields at 1 and not 0. This didn't used to be the case and was the cause of a lot of confusion!)

Table 5.4 summarizes the field ordering for 525-line and 625-line video.

Table 5.4 Video Field Ordering

Format	Field Topness/Ordering
525-line	Field 2, lower field first, even
625-line	Field 1, upper field first, odd

Video Line Sampling and DV

Unfortunately, when the DV format was developed, the format designers did not follow the ITU-R BT.656 specification. Though each line of video is sampled as specified in accordance with ITU-R BT.601 (Y'_{601}, C_B, and C_R sampling at 13.5 MHz with 720 Y' samples per line), they did not follow the specification for the actual lines to be sampled. Indeed, different decisions were made for the 525-line and 625-line formats.

525-Line DV

In 525-line they decided to use a 480-line active frame instead of a 486-line active frame. This decision was primarily due to the fact that DV uses a DCT (discrete cosine transform) compression that breaks the frame into 8×8 sample blocks. As 486 is not evenly divisible by 8, they decided to use only 480, which is evenly divisible by 8.

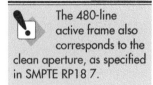

The 480-line active frame also corresponds to the clean aperture, as specified in SMPTE RP18 7.

Table 5.5 lists the line sampling used in DV 525-line formats. The standard 525-line digital line sampling is also provided for comparison.

Table 5.5 525-Line DV Format Sampling

Sampling Region	DV 525-Line	BT.656 525-Line
Field 1: start of active picture	23	21
Field 1: end of active picture	262	263
Field 2: start of active picture	285	283
Field 2: end of active picture	524	525

Notice that each DV field begins two lines later and ends one field earlier (e.g., line 23 versus 21, line 262 versus 263). This ensures that the field ordering for DV media is the same as that for other types of digital 525-line media.

625-Line DV

Unfortunately, though 625-line DV uses a 576-line active frame (as 576 is evenly divisible by 8), the lines sampled are not the same as those specified in ITU-R BT.656. This was due to incorrect assumptions made by the engineers who developed the DV format. Perhaps apocryphally, when the DV format was presented to both SMPTE and the ITU, this line sampling error was pointed out, but the manufacturers plainly stated that cameras and decks were already in production and the DV standard would not be modified to correct the error.

Table 5.6 lists the line sampling used in DV 625-line formats.

Table 5.6 625-Line DV Format Sampling

Sampling Region	DV 625-Line	BT.656 625-Line
Field 1: start of active picture	23	23
Field 1: end of active picture	310	310
Field 2: start of active picture	335	336
Field 2: end of active picture	622	623

Though field 1 is properly sampled, field 2 is offset upward by one line. This has the unfortunate side effect of changing the topness of the frame from field 1 to field 2 and therefore inverts the field ordering versus regular 625-line video. Fortunately, the Avid editing system properly handles mixing these two resolutions in the timeline but this difference can cause definite problems when you are exporting QuickTime® movies or other file-based video out of the system. Most other systems do not handle, or in some cases even understand, field order switching in the middle of a movie or sequence.

Table 5.7 Video Field Ordering

Format	Field Topness/Ordering
525-line	Field 2, lower field first, even
625-line (BT.656)	Field 1, upper field first, odd
625-line (DV)	Field 2, lower field first, even

Table 5.7 summarizes the field ordering for all 525-line and 625-line video formats.

6

THE WILD WORLD OF HIGH DEFINITION

"A camel is a horse designed by a committee."

—Sir Alec Issigonis

A good friend of mine, when faced with his first high definition (HD) online, wondered to me how a system (NTSC) that worked so well could have been turned into the nightmare of incompatibility that is high definition. Perhaps the best way to understand it is to realize that the NTSC format was a series of brilliant inventions and hacks that created a system that, despite its quirks, worked reliably and predictably. After it was up and running, a committee codified how it worked so others could understand it (via SMPTE 170M, and so on). Digital standard definition (SD) was therefore an attempt to bring this simple system into the digital age.

But HD was another thing altogether and in many respects truly is the metaphorical camel. By this I mean no disrespect to those who worked on the development of HD, but truly it is a system that attempts to satisfy many different goals, even if those goals are in opposition to one another. As with the camel, much of the design is absolutely brilliant and every part has its underlying key purpose. But the sum of the parts can boggle the mind. Let's begin by taking at look at the genesis of the HD format.

A Brief History of High Definition

Since the earliest days of video, engineers have always looked toward formats with higher quality and resolution. Indeed, the current NTSC format was called "high definition" at its time of development since it had more than double the scan lines of earlier experimental systems.

The genesis of the modern high-definition system began in 1968 when Japanese broadcaster NHK began work on a format called NHK Hi-vision. NHK Hi-vision was a 1125-line analog format that used a hybrid of both analog and digital compression to reduce the bandwidth requirements. The format was eventually named MUSE and went online in the early 1980s. MUSE used 1035 active interlaced scan lines and had an aspect ratio of 1.66:1. (For reference, NTSC is a 525-line analog format with 486 active scan lines while PAL is a 625-line analog format with 576 active scan lines.)

Around the time that the MUSE system went on-air, the Federal Communications Commission (FCC) began soliciting proposals for a next-generation video system. A number of companies and organizations put forward their own, often incompatible, format proposals. After years of hearing competing proposals and political arm-twisting, the FCC asked the groups to pool their resources and in 1993 the Grand Alliance was formed.

Prior to the formation of the Grand Alliance, there were 23 different format proposals made. These were eventually whittled down to 4 digital and 2 analog systems. The Grand Alliance focused not only on the high-definition digital video format, but also on the method of over-the-air transmission. As this book is focused on editing, we'll leave the discussion of over-the-air transmission to another book, such as *How Video Works*, Second Edition, by Diana Weynand and Marcus Weise (Focal Press, 2007).

The Grand Alliance members were AT&T, General Instrument Corporation, MIT, Phillips Consumer Electronics, the David Sarnoff Research Center, Thompson Consumer Electronics, and Zenith Electronics.

The Advanced Television Standards Committee

Out of the Grand Alliance was formed the Advanced Television Standards Committee (ATSC). Though the committee eventually decided on a single transmission format, as is often the case with committee-based standards, they didn't propose a single HD format, but instead released a list of supported video formats in what has become known as ATSC Table 3. They also agreed to use the MPEG-2 compression format for all signals.

The original ATSC Table 3 is produced here as Table 6.1 for your reference. Don't worry if this table generates more head scratching than useful information. That is the way virtually everyone feels when they first see it. We'll break the table's information down in a moment.

According to the ATSC format, as used in the United States, any of the ATSC Table 3 formats can be broadcast at the whim of the broadcaster and every high-defintion television (HDTV) set sold must support all of the listed formats. In reality, the broadcasters settled on two primary broadcast formats: one with 1080 active scan lines and one with 720 active scan lines. Other formats are used primarily for acquisition and mastering. The 480-line formats,

Table 6.1 ATSC Table 3 Compression Format Constraints

Vertical Size Value	Horizontal Size Value	Aspect Ratio Information	Frame Code Rate	Progressive Sequence
1080	1920	1, 3	1, 2, 3, 4, 5	1
			4, 5, 6, 7, 8	0
720	1280	1, 3	1, 2, 3, 4, 5, 6, 7, 8	1
480	704	2, 3	1, 2, 4, 5, 7, 8	1
			4, 5	0
	640	1, 2	1, 2, 4, 5, 7, 8	1
			4, 5	0

Legend	Horizontal Size Value
Aspect ratio information:	1 = square samples; 2 = 4:3 display aspect ratio; 3 = 16 × 9 display aspect ratio
Frame code rate:	1 = 23.976 Hz; 2 = 24 Hz; 3 = 25 Hz; 4 = 29.97 Hz; 5 = 30 Hz; 6 = 50 Hz; 7 = 59.94 Hz; 8 = 60 Hz
Progressive sequence:	0 = interlaced scan; 1 = progressive scan

Note: ATSC Table 3 does not include any 25- or 50-Hz formats. These were added later to support European broadcasters.

one of which was used by the Fox network for their early digital broadcasting, have fallen by the wayside.

In Europe, most digital broadcasts today are a digital PAL format with 576 active scan lines, similar to the 480-line formats used briefly in the United States. These are expected to transition to true HD broadcasts in the coming years.

1080-Line High Definition

This format is based on a video frame that contains a total of 1125 lines, 1080 of which are considered active. The 1080-line format is codified in SMPTE 274M. There are a total of 1920 active samples per line.

The 1080-line format includes 11 different subformats, or systems, each with its own specific frame rates and types (progressive or interlaced). Both R'G'B' and Y', C_B, and C_R (Y'$C_B C_R$) signals are supported. However, as R'G'B' signals are not yet supported by Media Composer, we will not discuss them in this book.

Though we refer to SD formats by the total number of scan lines (e.g., 525-line), HD formats are instead referred to by the number of active number of scan lines.

Supported Frame Rates and Types

Table 6.2 lists the 11 different systems specified in SMPTE 274M. We'll return to this table and add additional information throughout this section.

Table 6.2 SMPTE 274M 1080-Line Systems

System Number	System Name	Frame Type	Frame Rate (Hz)
1	1080p/60	Progressive	60
2	1080p/59.94	Progressive	59.94 (60 ÷ 1.001)
3	1080p/50	Progressive	50
4	1080i/60	Interlaced	30
5	**1080i/59.94**	**Interlaced**	**29.97**
6	**1080i/50**	**Interlaced**	**25**
7	1080p/30	Progressive	30
8	**1080p/29.97**	**Progressive**	**29.97**
9	**1080p/25**	**Progressive**	**25**
10	**1080p/24**	**Progressive**	**24**
11	**1080p/23.976**	**Progressive**	**23.976 (24 ÷ 1.001)**

Note: The systems listed in bold are currently supported by Avid Media Composer 3.0, but 1080p/29.97 is not available on software-only or Adrenaline-attached systems due to a hardware limitation.

Digital Component Sampling

As with SD component digital video, HD component digital video stores the video components Y', C_B, and C_R as three separate signals. The digital derivation of Y' from $R'G'B'$ uses a different weighted function from that used in SD component digital and analog video:

$$Y'_{709} = 0.7152\ G' + 0.2126\ R' + 0.0722\ B'$$

The Y' component is subscripted with 709 to indicate that the luma component is derived using the values ascribed in ITU-R BT.709, the international specification for HD digital video. The differences between the 601 and 709 derivations for luma are primarily due to the realities of modern television tube display capabilities. Indeed, it could be argued that tubes that generate a picture in accordance with the original NTSC or PAL standards never reached mass production. This was primarily due to difficulties manufacturing a stable green phosphor that corresponded to the original specification.

Digital Sampling Frequency

The 1080-line format uses 4:2:2 sampling with a luma sampling rate of 74.25 MHz. (You might recall that SD digital video uses a sampling rate of 13.5 MHz.) This sampling rate is used by the majority of the supported systems including the 60/30-Hz and

50/25-Hz systems. However, if you recall from Chapter 5, the NTSC format reduced the frame rate from 30 fps to 29.97 fps using a factor of 1.001.

To maintain compatibility, and more importantly timing, with simultaneous NTSC broadcasts, systems had to be created with a similar reduction. In these systems the 74.25-MHz sampling rate is divided by 1.001, the same factor used by NTSC to reduce the frame rate. (This rate is typically referred to as 74.25/1.001 MHz.) In addition, the high-rate progressive formats (1080p/60, 1080p/59.94, and 1080p/50) store twice as much information as the related interlaced formats (1080i/60, etc.) and therefore sample at double the 74.25-MHz rate, or at 148.5 MHz (or 148.5/1.001 MHz for 1080p/59.94).

All 1080-line formats have an identical number of active samples per line (1920). However, as was the case with ITU-R BT.601 and the 525-line and 625-line formats, the total number of samples per line varies from one frame rate to another.

Table 6.3 expands on Table 6.2 and includes the sampling frequency and total number of samples per line for each system. As with Table 6.2, the systems currently supported by Avid Media Composer are listed in bold.

Table 6.3 SMPTE 274M 1080-Line Systems

System Number	System Name	Frame Type	Frame Rate (Hz)	Sampling Frequency	Total Samples per Line
1	1080p/60	p (1:1)	60	148.5	2200
2	1080p/59.94	p (1:1)	59.94	148.5/1.001	2200
3	1080p/50	p (1:1)	50	148.5	2640
4	1080i/60	i (2:1)	30	74.25	2200
5	**1080i/59.94**	**i (2:1)**	**29.97**	**74.25/1.001**	**2200**
6	**1080i/50**	**i (2:1)**	**25**	**74.25**	**2640**
7	1080p/30	p (1:1)	30	74.25	2200
8	**1080p/29.97**	**p (1:1)**	**29.97**	**74.25/1.001**	**2200**
9	**1080p/25**	**p (1:1)**	**25**	**74.25**	**2640**
10	**1080p/24**	**p (1:1)**	**24**	**74.25**	**2750**
11	**1080p/23.976**	**p (1:1)**	**23.976**	**74.25/1.001**	**2750**

Note: Some manufacturer's documentation, most specific in reference to sync generators, uses the term 1:1 for progressive signals and the term 2:1 for interlace signals.

Voltage Sampling

Just as is the case with SD video, 1080-line systems use the voltage range specified in SMPTE N10. For Y' signals, black is assigned a voltage of 0 mV and white a voltage of 700 mV. The C_B and C_R signals use a voltage range between −350 mV and 300 mV. This sampling is shown in the following illustrations.

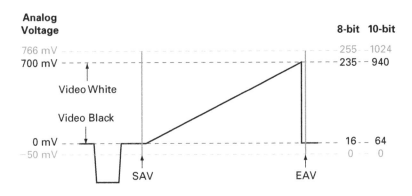

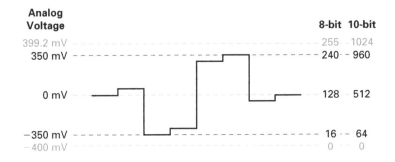

Signal Synchronization

If you recall from Chapter 5, SD signals use an analog sync pulse to synchronize two or more pieces of video equipment. This synchronization signal is displayed in the next figure for reference.

This type of sync signal is referred to as *bi-level* sync as the pulse has two voltages, a nominal voltage and a low voltage. In these systems, sync is triggered by the leading edge rise time. Bi-level sync's use of a low voltage adds a DC (direct current) component that, while not causing significant problems in low-bandwidth SD signals, introduces some synchronization complexities in high-bandwidth systems. In addition, generating a bi-level sync pulse in transmission actually requires a significant amount of power.

525-line video has a sync duration of 10.9 μs and an amplitude of −40 IRE. 625-line video has a sync duration of 12.0 μs and an amplitude of −300 mV.

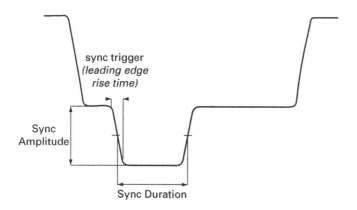

sync trigger
*(leading edge
rise time)*

Sync
Amplitude

Sync Duration

The developers of the HD format recognized that bi-level sync would not be sufficient for HD signals and instead used a *tri-level* sync. Tri-level sync has three distinct voltages instead of two and the rise time from negative to positive is used as the sync trigger as opposed to bi-level sync, which uses the leading edge of sync as the sync trigger.

The sync pulse begins at 0 mV, transitions to −300 mV for a specified duration, then transitions to +300 mV for the same duration, finally returning to 0 mV as shown in the following illustration.

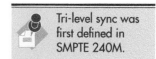

Tri-level sync was first defined in SMPTE 240M.

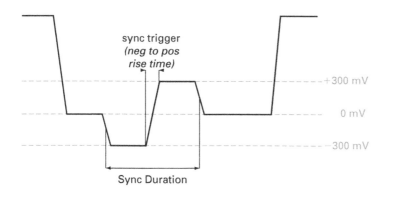

sync trigger
*(neg to pos
rise time)*

+300 mV

0 mV

300 mV

Sync Duration

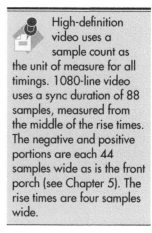

High-definition video uses a sample count as the unit of measure for all timings. 1080-line video uses a sync duration of 88 samples, measured from the middle of the rise times. The negative and positive portions are each 44 samples wide as is the front porch (see Chapter 5). The rise times are four samples wide.

The primary benefit of tri-level sync is that the symmetry of the sync signal results in a net DC value of 0 mV, eliminating the DC component introduced with bi-level sync. This makes signal processing and transmission far easier than with bi-level sync.

In all HD formats tri-level sync is applied to all three signals—Y′, C_B, and C_R—whereas it is only applied to Y′ in SD formats.

Unfortunately, whereas SD required only two different sync signals, one for 525-line video and one for 625-line video, the various frame rates and frame formats in HD mean that a separate sync generator is required for every single 1080-line format or system. Multiformat tri-level sync generators, such as the Tektronix TG700, can generate most of the tri-level signals, but few tri-level sync generators generate all of them. Fortunately, the majority generates sync for all of the more popular systems.

In addition, a standard NTSC black-burst generator can be used for the 1080i/59.94 format and a standard PAL black-burst generator can be used for the 1080i/50 format. It is still preferred to use tri-level sync for these formats, though.

As you can imagine, it is critical that the proper sync be provided to all decks and to the Avid system. Applying the wrong sync signal can result in signal distortions or even the inability to play, capture, or output a signal.

Line Structure

As mentioned earlier, 1080-line video has 1125 total lines. Similar to SD digital video, these lines consist of an active picture section surrounded by vertical blanking (or VANC) sections. The line structure is different for interlaced and progressive signals.

1080i Line Structure

Dividing the 1080-line format's 1125 lines by 2 results in a fractional result. This would imply that half-lines of picture and blanking are used in HD, just as they are in SD. As HD is a purely digital system, half lines are not required for synchronization and only full lines are used. Therefore, field 1 is comprised of 563 lines and field 2 is comprised of 562 lines.

Table 6.4 lists the interlaced line structure for 1080-line systems.

> If an HD signal is transmitted by an analog interface (such as the component analog outputs) half-lines are used and the two fields have 562.5 lines each.

Table 6.4 SMPTE 274M 1080-Line Structure

Field	Region	Lines
1	Vertical blanking	1–20
1	Active picture	21–560
1	Vertical blanking	561–563
2	Vertical blanking	564–583
2	Active picture	584–1123
2	Vertical blanking	1124–1125

Now let's look at the structure of the 1080-line interlaced frame in the following figure. Notice that the 1080-line format is field 1 ordered. This is true for all 1080-line systems, regardless of frame rate.

1080-line Interlaced Frame

f1 / 21 ———————————————	*first line of active picture*
f2 / 584 – – – – – – – – – – – – – – – – – – –·	
f1 / 22 ———————————————	
f2 / 585 – – – – – – – – – – – – – – – – – – –·	
⋮	
f1 / 559 ———————————————	
f2 / 1122 – – – – – – – – – – – – – – – – – – –·	
f1 / 560 ———————————————	
f2 / 1123 – – – – – – – – – – – – – – – – – – –·	*last line of active picture*

1080p Line Structure

Progressive video can be stored using two different approaches:

- *Progressive Segmented Frame (PsF):* This stores the progressive frame just like interlaced video by dividing the frame into two "fields," or segments. This storage method provides for better compatibility on some types of digital tape formats and is used, for example, by the HDCAM™ format developed by Sony. The two segments are reassembled on output to recreate the original progressive frame.
- *Progressive:* This stores the progressive frame as a single unit and does not subdivide it. Avid editing systems store and transmit progressive formats using this structure.

> Progressive segmented frame formats use the same line structure as 1080 interlaced.

Table 6.5 lists the progressive line structure for 1080-line systems.

Table 6.5 SMPTE 274M 1080p Line Structure

Region	Lines
Vertical blanking	1–41
Active picture	42–1121
Vertical blanking	1122–1125

720-Line High Definition

This format is based on a progressive video frame that includes a total of 750 lines, 720 of which are considered active. The 720-line format is codified in SMPTE 296M. There are a total of 1280 active samples per line.

The 720-line format includes eight different subformats, or systems, each with its own specific frame rate. Interlaced video is not supported by this format. As with the 1080-line format, both R′G′B′ and $Y'C_BC_R$ signals are supported by the standard, but only $Y'C_BC_R$ signals are supported by Media Composer at this time.

Supported Frame Rates and Types

Table 6.6 lists the eight different systems specified in SMPTE 296M. We'll return to this table and add additional information throughout this section.

Table 6.6 SMPTE 296M 720-Line Systems

System No.	System Name	Frame Rate (Hz)
1	720p/60	60
2	**720p/59.94**	**59.94**
3	**720p/50**	**50**
4	720p/30	30
5	**720p/29.97**	**29.97**
6	**720p/25**	**25**
7	720p/24	24
8	**720p/23.976**	**23.976**

Note: The systems listed in bold are currently supported by Avid Media Composer. Support for 720p/50, 720p/29.97, and 720p/25 were added in version 3.0.

Component Digital Sampling

As with the 1080-line format, 720-line video stores the video components Y′, C_B, and C_R as three separate signals and uses the ITU-R BT.709 digital derivation of Y′:

$$Y'_{709} = 0.7152\,G' + 0.2126\,R' + 0.0722\,B'$$

Digital Sampling Frequency

The 720-line format uses 4:2:2 sampling with a luma sampling rate of 74.25 MHz, just as is used in the 1080-line format. And, just as is done in the 1080-line format, the systems with frame rates of 59.94, 29.97, or 23.976 use a sampling rate of 74.25/1.001 MHz.

All 720-line formats have an identical number of active samples per line (1280). However, as was the case with the 1080-line formats, the total number of samples per line varies from one system to another.

Table 6.7 expands on Table 6.6 and includes the sampling frequency and total number of samples per line for each system. As with Table 6.6, the systems currently supported by Media Composer are listed in bold.

Table 6.7 SMPTE 296M 720-Line Systems

System Number	System Name	Frame Rate (Hz)	Sampling Frequency	Total Samples per Line
1	720p/60	60	74.25	1650
2	**720p/59.94**	**59.94**	**74.25/1.001**	**1650**
3	**720p/50**	**50**	**74.25**	**1980**
4	720p/30	30	74.25	3300
5	**720p/29.97**	**29.97**	**74.25/1.001**	**3300**
6	**720p/25**	**25**	**74.25**	**3960**
7	720p/24	24	74.25	4125
8	**720p/23.976**	**23.976**	**74.25**	**4125**

Voltage Sampling

Just as is the case with 1080-line systems, 720-line systems use the voltage range specified in SMPTE N10. For Y' signals, black is assigned a voltage of 0 mV and white a voltage of 700 mV. The C_B and C_R signals use a voltage range between −350 mV and 300 mV.

Signal Synchronization

The 720-line formats use tri-level sync, as is used for 1080-line formats. The sync pulse uses the voltages 0 mV, −300 mV, and +300 mV with the negative to positive rise time used as the sync trigger. The sync signal for 720-line video measures 80 samples wide, as measured from the middle of the rise times. The negative and positive portions are each 40 samples wide. The width of the front porch varies from system to system.

A specific sync generator is required for all 720-line systems. A standard NTSC or PAL black-burst cannot be used for any 720-line format.

Line Structure

As mentioned earlier, 720-line video has 750 total lines, consisting of an active picture section surrounded by vertical blanking (or VANC) sections. Table 6.8 lists the interlaced line structure for 720-line systems.

Table 6.8 SMPTE 296M 720p Line Structure

Region	Lines
Vertical blanking	1–25
Active picture	26–745
Vertical blanking	746–750

Working with High Definition in Avid

Since they first supported HD projects, Avid editing systems have been able to freely mix and match SD and HD material in a sequence as long as the *frame* rate of the two formats matched. Version 3.0 adds the ability to mix not just SD and HD in a sequence, but all HD formats that share a common timeline. This includes both progressive and interlaced formats. For example, that means that a single sequence can include the following formats: NTSC (29.97 fps), 1080i/59.94, 1080p/29.97, and 720p/29.97. Note, however, that you cannot mix 720p/59.94 as it has double the frame rate (59.94) as the other formats. We certainly hope to see this capability appear in a future release, but we won't include it until we are able to do so in real time with a level of quality equivalent to a high-quality outboard standards converter.

When working with a mixed-format sequence, you may need to switch between the different formats supported. This is accomplished via the Project Type tab located in the Format tab of the Project window. All formats compatible with your current project format will be listed. (We'll talk more about compatible project formats in Chapter 9 when we discuss conforming and finishing strategies.)

Subsampled High-Definition Rasters

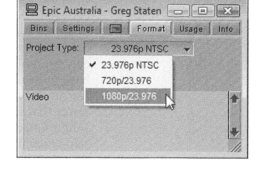

Though DNxHD media uses the full HD raster (1920 or 1280 samples), most HD camera formats do not. Though you'll capture some of these via HD-SDI (also called baseband) and the camera/ deck will resize the raster to full width, if you are bringing in a file-based format such as XDCAM HD or P2 or capturing via FireWire either DVCPRO HD or HDV, you will be capturing media that actually uses a subsampled, or *thin*, raster. Camera formats often use a subsampled raster for a number of reasons including space consumption and to match the actual imaging sensor. Table 6.9 lists the raster sizes used by popular file and FireWire formats. Note that some formats only use a "thin" raster for 1080-line systems, not 720-line systems. In those instances we have only included the 1080-line system.

Table 6.9 "Thin" Raster Camera Formats

Camera Format	HD System	Raster Size
HDV	1080-line	1440 × 1080
DVCPRO HD	1080-line	1280 × 1080
	720-line	960 × 720
XDCAM HD (18, 25, 35 Mbit only*)	1080-line	1440 × 1080
XDCAM EX (SP mode only*)	1080-line	1440 × 1080

*Both XDCAM HD and XDCAM EX support both thin and full-width rasters. To use the full-width raster in XDCAM HD, select 50-Mbit recording (only available on some XDCAM HD cameras and decks). To use the full-width raster in XDCAM EX, select 35-Mbit HQ recording.

Avid Media Composer 3.0 natively supports the HDV, DVCPRO HD, and XDCAM HD/EX raster sizes, though not on every hardware configuration. The Nitris hardware (used in Avid Symphony Nitris) does not support thin rasters of any type and therefore these rasters are missing. The Adrenaline DNxcel HD hardware only supports the full raster and the HDV (1440 width) raster. Fortunately, this is the same raster size used by XDCAM HD and XDCAM EX so you can use this option for those formats. The DVCPRO HD raster, though, is not available. The new Mojo DX and Nitris DX hardware support all thin rasters and will resize them on-the-fly in hardware to full width for baseband output to a monitor or deck.

As is the case for compatible HD formats, you can freely switch between the available raster types in a project. Keep in mind, though, that switching raster types will definitely affect your real-time effect playback performance. Why? Well, quite simply, if the clip you are playing does not match the raster the system will have to resize the frame on-the-fly to the selected raster. For this reason, your best performance will always come from using the raster that matches your material. If you have mixed rasters in your sequence (e.g., standard DNxHD and XDCAM HD in the same timeline), use either the raster that matches the majority of your footage, or the thinnest raster, as a computer can always resize down (decimate) faster than it can resize up (extrapolate).

Original Image Prior to Correction

Corrected Grayscale Image

Color Figure 1

Original Image Prior to Correction

Corrected Grayscale Image

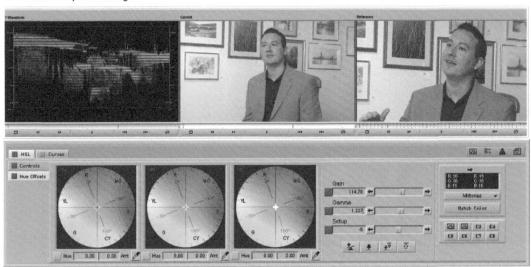

Color Figure 2

Original Image Prior to Correction

Corrected Grayscale Image

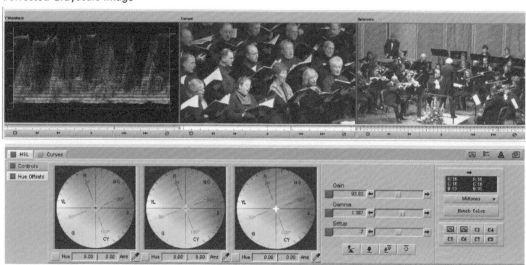

Color Figure 3

Original Image Prior to Correction

Corrected Image

Color Figure 4

Original Image Prior to Correction

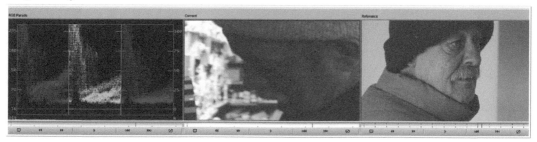

Corrected Image

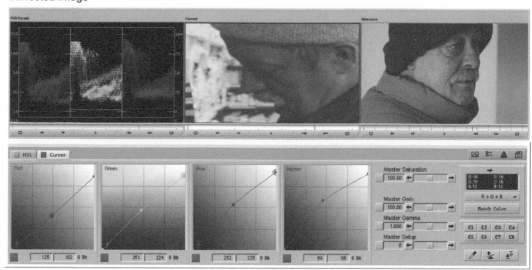

Color Figure 5

Original Image Prior to Correction

Corrected Image

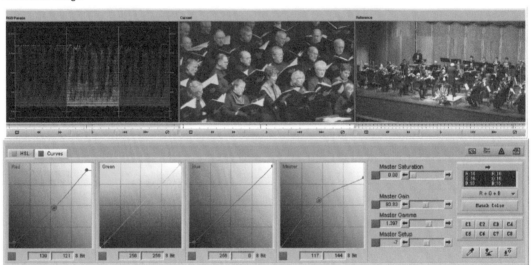

Color Figure 6

Original Image Prior to Treatment

Treated Image

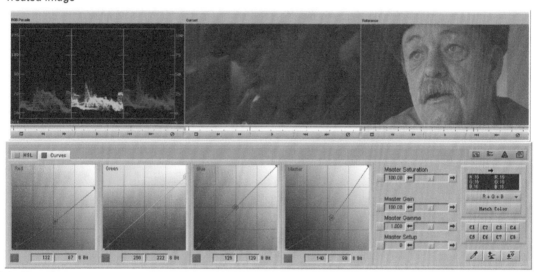

Color Figure 7

Original Image Prior to Treatment

Treated Image

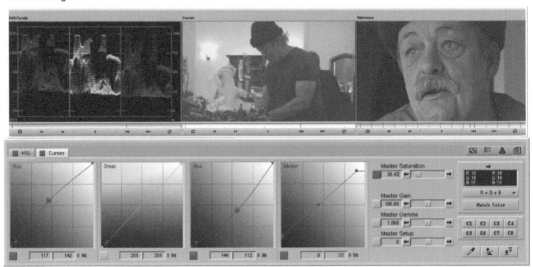

Color Figure 8

7

IMPORTING AND EXPORTING

"Editing is a natural extension of collage making."

—Rachel True

Graphics come in many shapes and sizes from artists, ad agencies, the Internet, scanners, digital cameras, and a wide range of other graphics programs, but very rarely will they be perfectly prepared for video. In general, graphics tend to be the wrong size or the wrong resolution either through accident, ignorance, or repurposing. This chapter will help you learn how to compensate for that.

Avid has built in some system intelligence to deal with all the different graphic formats and, if the graphic format given to you is recognized, the system automatically imports it. As opposed to many programs, a file extension is not required for import—a very beneficial fact if you are editing on a Windows-based Avid system and working with graphic artists who use Macintosh systems. Despite this fact, it is still a good idea to append the correct file extension to each graphic in case you need to fix a problem in Adobe Photoshop® or some other program, as few other programs on Windows can read a file that is missing the correct extension.

Also, the new drag-and-drop capabilities of the last few releases of Media Composer speed up workflow and simplify basic tasks. The user creates an export template based on the requirements outlined in this chapter and is assured every export will be consistently correct. By dragging a sequence or a master clip to the desktop level, even complicated exports can be done by beginners. The same HIIP (host image independence protocol) technology allows users to drag a graphic straight from the network and drop it on the open bin. By creating preset, named, copied and carried, and import and export templates, you can make interoperability with other software a one-step process.

Despite the fact the system supports 26 different import file formats, we strongly recommend that you use either the TIFF or

PNG format when creating graphics for import. PNG is especially useful when creating graphics with alpha channels as Adobe Photoshop automatically generates a straight alpha channel for the composite of all layers. In addition, PNG files do not support the CMYK (cyan, magenta, yellow, black) color space, eliminating one of the "gotchas" we'll discuss below.

Import and Export Basics

Some basic issues should be understood when working with computers, graphics, and video. Because of the way computers have developed, with their reliance on RGB (red, green, blue) color for their screens and memory, and the way video developed, beginning with black-and-white analog, the two mediums have never had a particularly easy coexistence. One of the hopes for high-definition television (HDTV) is that some of those issues will be resolved, but the addition of more incompatible formats has rarely made things simpler.

Color Space Conversion

Computers and video work with different color formats and different kinds of scanning and scan rates. We commonly refer to methods of representing colors as *color space*. It usually is represented graphically by a cube with white at the top and black at the bottom. The range of colors possible within a color space makes up the height and width of the cube. Different color spaces use different methods of distributing those colors inside the cube shape.

Computers traditionally work in RGB color space and digital component video works in Y', C_B, and C_R ($Y'C_BC_R$). RGB is easier for displaying images on an RGB computer screen and working with computer memory. RGB has a greater range of colors to choose from than video, especially in the yellow hues. Thus, a conversion between $Y'C_BC_R$ and RGB can force colors to change because a true exact match does not exist or because colors are out of the range in the new color space ("out of gamut").

Fortunately, most of the available spectrum in RGB *does* properly map to the $Y'C_BC_R$ video's color gamut, so if your colors or video levels are changing, it is more likely you are doing something wrong like exporting or importing into the system using the RGB levels choice instead of the 601 levels. This chapter will discuss the details of these choices later.

CMYK Color Space

The CMYK color space is specifically used for color-offset printing onto paper. This color space uses *subtractive* colors to remove color from the white paper and generate a full-color image. (In contrast, video monitors use the *additive* colors red, green, and blue to create color from black.) Images saved in the CMYK color space are not compatible with video applications and should be converted to RGB, using a program such as Adobe Photoshop, prior to bringing into your edit bay.

Square versus Nonsquare Pixels

When graphics and animations are created for use in Avid editing systems, they can be created using either square or nonsquare pixels. Standard-definition (SD) digital video uses nonsquare pixels, while high-definition (HD) video uses square pixels.

Virtually all computer display cards use square pixels. Because the display uses square pixels, most graphic and animation programs also use square pixels. With square pixels, a 100×100 pixel box would be a perfect square.

However, SD digital video does not use square pixels. Both the ITU-R BT.601 and DV digital video standards use a 720-pixel width for both NTSC and PAL. But, because NTSC and PAL have different numbers of scan lines (486 for ITU-R BT.601 or 480 for DV versus 576), SD digital video has pixels that are stretched vertically for NTSC and stretched horizontally for PAL.

The following graphic shows a close-up of a circle drawn with square pixels and NTSC and PAL nonsquare pixels. Notice that the square-pixel circle has the same number of pixels both horizontally and vertically, while the NTSC and PAL circles do not.

Computer Pixels
(Square)

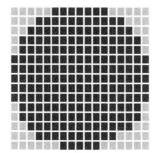

NTSC ITU-R BT.601 Pixels
(Non-Square)

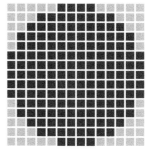

PAL ITU-R BT.601 Pixels
(Non-Square)

Because SD video uses nonsquare pixels, graphics are usually created at an intermediate square pixel size (e.g., 648×486 or 768×576) and then resized to 720×486 by the Avid editing system during import.

High-definition video, on the other hand, uses square pixels. Therefore, HD graphics can be created at the native size (1920 × 1080 or 1280 × 720), not an intermediate size as with SD.

Graphics and animations can be created with either square or nonsquare pixels. However, if the proper frame size is not used, the graphic or animation will be distorted when imported into the system. Table 7.1 lists the proper sizes for square and nonsquare pixel frames.

Table 7.1 Proper Square and Nonsquare Pixel Frame Sizes

Format	Square Pixel (4 × 3)	Square Pixel (16 × 9)	Nonsquare Pixel (4 × 3 and 16 × 9)
NTSC (601)	648 × 486	864 × 486	720 × 486
NTSC (DV)*	640 × 480	853 × 480	720 × 480
PAL	768 × 576	1050 × 576	720 × 576
1080-line HD	N/A	1920 × 1080	N/A
720-line HD	N/A	1280 × 720	N/A

*As mentioned in Chapter 5, NTSC DV does not use the full-frame ITU-R BT.601. Instead, it omits four lines from the top and two lines from the bottom of the frame. As a result, native DV graphics have a different frame size than regular NTSC.

The following guidelines should help you determine whether to use nonsquare or square pixels when importing and exporting SD frames and clips. Use nonsquare pixels when:

- Importing or exporting using the SD versions of the Avid QuickTime codec. The Avid QuickTime codecs for SD video require nonsquare pixels. These codecs are discussed in detail later in this chapter.
- Exporting SD video out of an Avid editing system. Because the 601 or DV frame is the native frame size for SD Avid editing systems, if you export using the proper nonsquare pixel size, there is no risk of artifacting due to a resize from nonsquare to square pixels.
- Creating SD animations and composites for import into Avid editing systems. You should always render animations and composites to the native frame size for the system into which you are importing those files. We will discuss this in greater detail later in the chapter.

Use square pixels when:

- Preparing an SD still graphic for import. Sizing to a square pixel of 4 × 3 or 16 × 9 aspect ratio is the simplest method and is appropriate for still graphics.

- Exporting an SD still graphic for use in print or on the Web. Any image you plan to export for use in print or on the Web should be at a square pixel size so it does not appear distorted when printed or displayed. You should also export only one field.
- Importing or exporting an HD frame or clip. As the HD frame uses square pixels natively you should always use square pixels for both import and export.

Voltages and Video Graphics

Avid editing systems allow you to import and export animation, video, and still images using either RGB levels or ITU-R BT.601 (for SD) or ITU-R BT.709 (for HD) levels. As an editor, you must understand the differences between the two choices and be able to communicate those differences to the people who are producing the import elements for your project.

When computer graphics are created, they are often created with absolute values for black and white. In 24-bit RGB (8 bits for each channel), black is assigned a value of 0 and white a value of 255. There is no allowance for values beyond either black or white.

However, the ITU-R BT.601 and ITU-R BT.709 digital video standards do not treat black and white as absolutes—excursions above white and below black are allowed. To maintain full compatibility, Avid systems allow the creation of graphics using either computer graphics mapping (often referred to as RGB mapping) or ITU-R BT.601/ITU-R BT.709 mapping (also referred to as 601/709 mapping). All Avid editing systems use 601/709 mapping internally. Let's take a look at the differences between the two mapping options.

RGB Mapping

RGB mapping assumes that video black (NTSC: 7.5 IRE, HD, and PAL: 0mV) is assigned a value of 0 and video white (NTSC: 100 IRE, HD, and PAL: 700mV) a value of 255. There is no allowance for excursions above these values. If an image is exported out of an Avid editing system using RGB mapping, any values below video black or above video white will be clipped. This results in the signal mapping as shown in the following illustration.

Most graphics and animation packages, including Adobe Photoshop and Adobe After Effects, assume RGB mapping. It is appropriate for graphics created for print and onscreen use, as black and white need to be absolute values. The concept of "whiter than white" or "blacker than black" does not come into play.

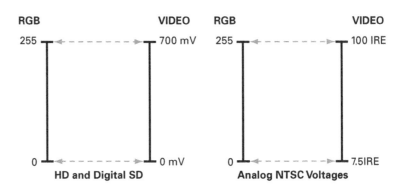

601/709 Mapping

Recall from Chapter 5 that the ITU-R BT.601 and ITU-R BT.709 standards use identical voltage-to-pixel sampling. The primary difference between the two is that the 709 standard more accurately reflects the color gamut that is produced by modern video monitor phosphors.

The ITU-R BT.601 and ITU-R BT.709 digital video standards allow for excursions beyond video black and video white. This ensures that some camera overexposure is maintained and allows for subblack values for luminance keying. The ITU-R BT.601 standard specifies that black is at 16 and white at 235. This allows for a reasonable amount of signal footroom and headroom and results in the signal mapping shown in the following illustration.

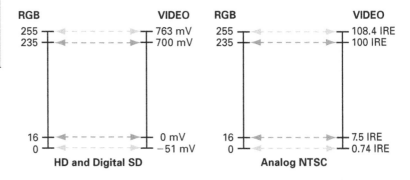

When a video signal is hard-clipped at video black and video white, as it is with RGB graphics mapping, undesirable "blooming" or flat regions often result. Additionally, slight "ringing" due to compression or analog filtering is often converted to blooming and therefore amplified. By using ITU-R BT.601 or ITU-R BT.709 mapping, you can eliminate or dramatically reduce both of these problems. This mapping also allows graphic artists to create true luma keys, since you can represent key black (a value blacker than black). As mentioned previously, Avid editing systems use 601/709 mapping internally. If you need to maintain all of the video signal information when you export a clip, you should use 601/709 mapping. However, not all third-party programs natively understand this mapping. Extra care might need to be taken by the graphic artist, animator, or compositor to make sure that the values for video black and video white are maintained and not allowed to extend into the headroom or footroom.

When importing graphics and animations, be sure to select the correct mapping. If the wrong mapping is chosen, the signal values will be incorrect. Table 7.2 describes what happens when the wrong mapping is chosen.

Table 7.2 RGB and 601/709 Mismatch Results

File Has	Imported As	Result
RGB values	601/709	Luma and chroma are stretched—image appears to have greater contrast. Video black lowered to −51 mV (0.74 IRE). Video white raised to 763 mV (108.4 IRE). Valid chroma might now be out of bounds.
601/709 values	RGB	Luma and chroma are squeezed—image appears to have lower contrast. Video black raised to 50 mV (14 IRE). Video white lowered to 640 mV (94 IRE).

Avid editing systems allow you to export and import graphics and animations using either RGB or 601/709 levels. The following guidelines should help you determine when to use each mapping.

Use 601/709 levels when:

- Exporting a frame or frames that you plan to modify subtly and reimport. This method is appropriate when you need to fix a dropout or touch up negative grit. Using 601/709 levels maintains all of the captured signal. If you use RGB levels, the system clips all values below video black and above video white, which might introduce undesirable artifacts and cause the modified frame not to match back in perfectly.
- Using or creating video that requires superblack, such as a luma key element.

Use RGB levels when:

- Exporting a frame or frames that you plan to modify radically and reimport. One example is when you need to apply a Stylize effect in Adobe Photoshop. Using RGB levels clips the signal at video black and video white, which is necessary in this case. If you use 601/709 levels, the effect you apply might cause the signal to extend beyond video black and video white.

- Exporting a frame to be used in print or on the Web.

Field Ordering

If the element (graphic or animation) to be imported has been field rendered or if it contains interlaced video, it is critical that the file has the proper field ordering. Field ordering defines how the frames within the file are interlaced.

- An odd, or upper-field, ordering uses the first line of each frame for field 1.
- An even, or lower-field, ordering uses the first line of each frame for field 2.

Whenever you create an animation or video composite in a third-party program for import into Avid, you must set the field ordering correctly or the file will not play back correctly once imported.

Table 7.3 lists the proper field ordering that should be used when creating animations for import or exporting video out of the Avid editing system.

Table 7.3 Proper Field Ordering for Import

NTSC	PAL 601	PAL DV*	HD
Even (lower field first)	Odd (upper field first)	Even (lower field first)	Odd (upper field first)

*Recall from Chapter 5 that PAL DV is different than PAL 601 due to incorrect line assignment in the DV standard.

Fields and Still Graphics

Another characteristic of graphics that must be taken into consideration is a video field. Broadcasting an interlaced signal takes less bandwidth because only half of the image is transmitted at any one time, but it complicates things when you are taking an interlaced image to a noninterlaced medium like the computer. A computer uses a progressive scan for display on your monitor, which means that the image is drawn on the screen as a single frame, not two fields. If you export an interlaced frame from the Avid editing application and there is some kind of horizontal motion in the frame, you see a difference between the first set of scan lines, field 1, and the second set, field 2. Although they are only a fiftieth or sixtieth of a second apart, you see jagged horizontal displacement of the image every other line. If you are working with interlaced images you will need to consider when and how to de-interlace them when exporting to graphics or animation programs.

With the ease of basic desktop editing and the combination of graphics that go straight from one computer graphic format to a computer video format, you have some quality challenges. All of these image type mismatches can be dealt with if you are careful when converting formats. These graphics have too much fine detail to reproduce well in the relatively low-resolution, interlaced world of SD video. Consequently, the images buzz, flicker, and give us unpredictable results if played back with little compression. A thin line may look fine on a progressive scan monitor, but the moment it moves to an interlaced scan medium like SD video, that line may be only one scan line wide. This means the line is drawn on the screen only every other field, causing a disturbing flicker. Images that originated with a video camera can never be recorded with that kind of problem. And since, in the past, most graphics were seen through video monitors as they were being created, they could be adjusted on the spot so that the design could take into consideration the limitations of what looked good on video. Colors were toned down and detail was blurred until the image was acceptable, and then it was put to tape.

Until all graphic workstations can figure out how to approximate what the final product will look like after being interlaced and reduced in resolution for SD broadcast, you must be able to tweak the graphics after you receive them. The most common adjustments are to open the graphic in a graphics program on the editing workstation and add a little blur to areas that are buzzing with too much detail. If done with a little skill, the blur will never be noticed. In fact, a very slightly blurred image looks better than one that is too sharp. Though you could use the Paint effect in Avid to blur the file, the best approach is to use a deflickering effect such as the Reduce Interlace Jitter effect available in Adobe After Effects®. Another common adjustment is to lower the saturation of a particular color or, in a worst-case scenario, all the colors. Again, this can be done easily and safely using the Avid color effect or a safe color-limiter effect.

Alpha Channels: Straight or Premultiplied?

Currently Avid editing systems do not support premultiplied alpha channels. It is critically important that all imported graphics and video that have an alpha channel be created with a straight alpha, and not a premultiplied alpha. Nearly all alpha channels created in Adobe Photoshop are straight alphas, but many programs, including Adobe After Effects, create premultipied alpha channels by default.

If you import a graphic created with a premultiplied alpha into an Avid editing system, there will be a black halo around the edges of the graphic. This halo cannot be removed within the Avid and can only be removed by re-rendering with a straight alpha or converting the alpha using Adobe After Effects.

Understanding Premultiplication

Premultiplication is a method by which the alpha channel is applied to the file's foreground in order to modify it. An easy way to understand premultiplication is to think of an alpha channel as a cookie cutter. Let's take a simple example where the file's foreground is a solid color and the alpha channel contains a logo. If the file was saved with a straight alpha, the foreground is left alone and only the alpha channel represents the logo's shape.

Original Graphic

**Premultiplied Version
Imported into Avid**

**Straight Version
Imported into Avid**

When a file is premultiplied, the alpha channel is applied to the foreground. This is similar to using a cookie cutter to cut a shape out of a sheet of dough. The surrounding "dough" is removed and replaced with a specific color, usually black.

Notice that the alpha channel is identical in both of the above images. Premultiplication does not affect the alpha, but instead affects the foreground.

Premultiplication is very easy to understand when the alpha channel is purely black and white with no intermediate grays.

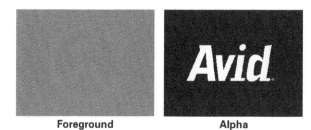

Foreground   Alpha

Now let's examine what happens when the alpha has gray, or partially transparent, areas. Imagine that our image is a blurred registration mark, as shown in the following illustration.

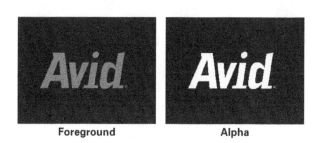

Foreground   Alpha

The blurred edge is partially transparent and, when composited against another image, will be blended with the other image. Now let's look at how a compositing program renders the foreground and the alpha channel.

If the image is saved with a straight alpha, the shape of the registration mark is expanded so that the color of the object exists for every pixel of the object. This includes all of the partially transparent pixels, even those that are barely visible. The foreground looks like it was cut out by a fat version of the alpha. The next illustration shows what the foreground and alpha look like after rendering.

Now let's look at the same image when rendered as a premultiplied alpha. In this case, when the alpha channel is applied to the foreground, the partially transparent areas are composited with a specific color, again usually black.

Working with Straight and Premultiplied Images

Because the foreground is stored very differently for straight and premultiplied images, it is critical that the image be interpreted properly or it won't composite correctly. Let's take a look at how a compositing program interprets straight and premultiplied images.

Foreground	Alpha

Straight Alpha (Not Premultiplied)

The compositing of straight alphas is very straightforward. The alpha channel is applied like a cookie cutter to the foreground and the surrounding information is ignored. Because the color of the foreground exists for both opaque and partially transparent pixels, the color of the foreground is preserved.

Premultiplied Alpha

Remember that when an alpha channel is premultiplied, the foreground is composited with black. If the foreground object was red, the partially transparent areas of the foreground were stored not as a pure red, but as a blending of red and black. Before the image can be composited against another image or a video clip, the black must be removed from the foreground. Compositing programs do this by applying an identical mathematical function to both the alpha channel and the foreground, in essence "unmultiplying" it.

Premultiplication Guidelines

Avid editing systems do not know how to correctly interpret premultiplied alphas. Therefore, it is critical that animations be created using straight alphas.

If a premultiplied alpha is imported into the Avid editing system, artifacts will be visible. To illustrate these artifacts, we will take our image of the blurred registration mark and composite it against a solid color.

Because a straight alpha is used purely as a cookie cutter to extract the shape from the foreground, the foreground pixels are extracted exactly as they appeared in the foreground. Because transparent pixels in a premultiplied alpha image have been blended with black, this results in a black halo around the object. (If the image had been premultiplied with white, a white halo would be visible instead.)

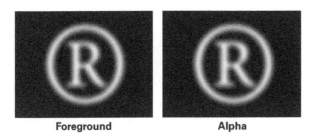

Foreground   Alpha

Always create animations that will be imported into Avid editing systems using straight alphas.

Sequential Files

Avid Symphony Nitris can import animation and video that is stored in a sequential file format. Unlike a QuickTime or AVI file where the entire animation is stored in a single file, the sequential file format stores each frame as its own file. The files are numbered to identify the frame order (e.g., open.000.tif, open.001.tif, open.002.tif, and so on). Sequential files can be stored in any still-frame format.

Configuring the Import Setting

When correctly importing a graphic, several important choices must be made. The dialog boxes are designed to reflect the type of graphic to be imported. The system will import graphics correctly, assuming that accurate information is given about your graphics. The next few sections will show you the differences between settings.

Aspect Ratio, Pixel Aspect: 601

- *601/709, nonsquare:* (System default.) Assumes the file is properly sized for import and leaves the image alone. If the image is not properly sized, this option forces the image to fit the entire video frame and will distort images with non-SD or non-HD television aspect ratios depending on your project format. This option should also be used when

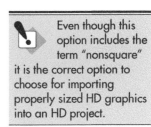

Even though this option includes the term "nonsquare" it is the correct option to choose for importing properly sized HD graphics into an HD project.

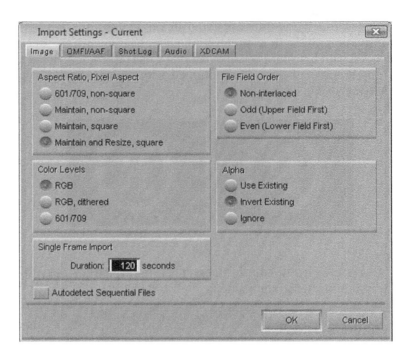

importing HD-formatted graphics into 16×9 SD projects or 16×9 SD-formatted graphics into HD projects.

- *Maintain, nonsquare:* (For use with NTSC projects only.) Designed to be used with nonsquare NTSC DV (or DVD) images imported into a standard NTSC resolution. NTSC DV has a frame size of 720×480. The 720×480 image is centered in the frame and video black is added at the top and the bottom to pad the image out to 486 scan lines. This option should also be chosen if importing a 720×486 frame into an NTSC DV resolution. In that case, the top four lines and bottom two lines of the 720×486 frame are removed from the image. This conforms to the SMPTE specification for NTSC DV frames.

- *Maintain, square:* Designed to be used with images that are smaller than the video frame size and cannot be resized. It does not attempt to resize the image, but compensates for the square pixels, centers it within the video frame, and adds video black around the image. This option is designed to make it easy to bring in small graphics, such as Web-originated art, into the Avid editing system. If you import larger than expected graphics using this option, the graphic will be resized and, if necessary, letterboxed. If you import SD square pixel graphics (e.g., 648×486 or 864×486) into an HD project using this option, the graphics will be centered in

If the graphic has an alpha channel and one of the three Maintain options is chosen, the system will key out the area around the graphic instead of adding video black.

the video frame and will not be resized. This is often the preferred method to bring in SD graphics with alpha channels into an HD project.

- *Maintain and Resize, square:* Assumes an incorrect image size. It letterboxes the image with video black and resizes it to fit either the maximum width (for wide images) or height (for tall images). It also assumes the import file has square pixels and compensates accordingly. If you import larger than expected graphics using this option, the graphic will be resized and, if necessary, letterboxed. This option can be used when importing HD-formatted graphics into a 4×3 SD project or 4×3 SD-formatted graphics into an HD project. In either case, the graphics will be resized to fit the video frame and letterboxed or pillarboxed.

For reference, Table 7.4 summarizes the correct aspect ratio, pixel aspect options for properly formatted SD and HD graphics.

Table 7.4 Aspect Ratio, Pixel Aspect Options for SD and HD Graphics

Source Format	Import into 4×3	Import into 16×9 SD	Import into 16×9 HD
4×3 SD (e.g., 648×486)	601/709, Nonsquare	Not supported	Maintain and resize, Square
16×9 SD (e.g., 864×486)	Maintain and resize, Square	601/709, Nonsquare	601/709, Nonsquare*
16×9 HD (e.g., 1920×1080)	Maintain and resize, Square	601/709, Nonsquare	601/709, Nonsquare

*Will resize the image to the full height of the HD frame. If you want to import the graphic without resizing, use "Maintain, Square" instead.

File Field Order

File field order allows you to set the field ordering of the imported file. If the file is not interlaced (frame rendered), set this option to "Noninterlaced." Otherwise, refer to Table 7.5 to choose the correct option.

Table 7.5 Proper Field Ordering for Import

NTSC	PAL 601	PAL DV*	HD
Even (lower field first)	Odd (upper field first)	Even (lower field first)	Odd (Upper field first)

If you are importing a QuickTime movie that is encoded with an Avid QuickTime codec, this setting is ignored and the animation's field ordering cannot be changed. Therefore, it is critical that all animations and composites are rendered with the correct field ordering.

Color Levels

Before using RGB, dithered, you might want to try reimporting graphics that display with banding using a 10-bit resolution.

- *RGB:* This option is designed to be used with traditionally created computer images. The blackest black in the graphic will be assigned the value of video black and the whitest white will be assigned the value of video white. This option should be chosen for graphics and animations created in third-party programs unless the graphic or animation uses 601 levels.
- *RGB, dithered:* Assigns values identical to RGB. Select this option if you are importing a graphic with a fine gradient. Due to the limitations of 8-bit 4:2:2 video encoding, banding is possible in fine gradients. This option adds a slight amount of noise to the gradient and can mask the banding inherent in digital video.
- *601/709:* Use this option if the graphic was created specifically to use the extended signal range available in either the ITU-R BT.601 (SD) or ITU-R BT.709 (HD) video standards. Do not use this option if the graphic was not created for 601 import as illegal color values may result.

Alpha

- *Use Existing:* Applies only to images that have an alpha channel; the setting has no effect on images that don't have an alpha channel. Use this option when importing movies rendered with the Avid QuickTime codec.
- *Invert Existing:* Inverts the black areas and white areas in an alpha channel. Use this option when importing still graphics, sequential file animations or movies, or movie files created with standard codecs.
- *Ignore:* If this option is selected, the system disregards the alpha channel and imports only the RGB portion of the image.

Single Frame Import

Use this option to set the duration, if desired, of an imported still graphic. This option sets the maximum duration for an imported still. Once imported, the graphic cannot be trimmed out beyond this duration. I strongly recommend setting it to a long duration if you plan to have your graphic onscreen for a significant duration. Otherwise you will have to edit it in multiple times. Regardless of the duration you choose, the media will only take up one frame of space on disk.

Autodetect Sequential Files

This option, off by default in Media Composer 3.0, tells the system whether or not to look for a numbered sequence of files and import them as a single file. This option can backfire on you if you have graphics in a file named "Logo version 1," "Logo version 2," and so on. If enabled it would import both of those files as a single file with each file having a duration of one frame. If that isn't what you want make sure this option is disabled!

In addition, in some versions of the Macintosh operating system, enabling this option will actually *hide* folders and files that are sequentially numbered. If you can't find the file you're looking for, try disabling this option.

Exporting

There is certainly a joy to working on a general-purpose computer instead of a dedicated graphics workstation. You always have the ability to quickly export a frame from the video application, tweak it in a graphics program, and import it back. You no longer need to call the graphics department at the last minute to make the minor changes that are inevitable as the deadlines get closer. It makes it easy to take any frame and use it as a background, do some simple rotoscoping (painting on the video frame by frame), or isolate parts of a frame with an alpha channel. The trap, of course, is that you will always be counted on to do this once you have shown how easy it is!

Export Templates

Since exporting usually is done with only a handful of the potential formats, most people find the export formats that suit them and ignore the rest. There is also a pattern to the type of exports; users find the video format or the graphic format they prefer and stick with it. It makes sense then to take the settings that are used in exporting most often and save them as templates. You can create and save these templates as user settings so you can take them with you from job to job with the assurance that you will always get the export right if you make the templates right. It is also great for experienced editors to make them for their less-experienced colleagues or assistants, who can use them with an extra level of confidence knowing that they will be correct.

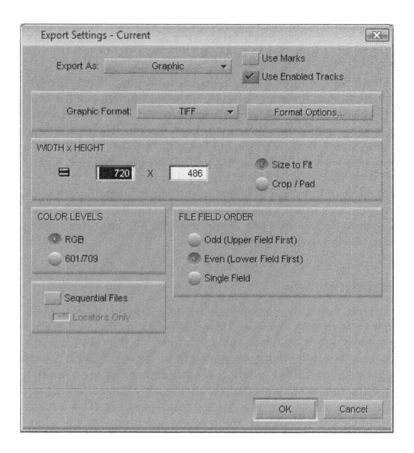

Export Basics

Exporting has its own set of choices, but most of them are decided by the ultimate use of the image. If you are exporting an image to work on and then you re-import back, you want the export and import to be exactly the same. This means choosing 601 video levels and the native frame size. In NTSC that native frame size is 720 × 486 and in PAL it is 720 × 576.

If you change the frame size during export, two things happen. First, it takes much longer to export since the application must do more work resizing. Second, and more seriously, the scan lines are disturbed from their precise, standardized relationship. If you export a still at a small size for use in a document or for a Web page, the scan lines are not as important because the image is not going back to full-frame video. A video imported back to the Avid system needs the interlaced scan line information to reproduce the image exactly. This means you should not de-interlace the image in the graphics program if you are going to re-import it to the Avid system. If you resize the image, even to a square pixel size like 720 × 540 or 768 × 576, when you bring the image back to a nonsquare video playback like on the Avid system, the import process does not know exactly how the original scan lines were laid out. The image will be degraded.

During import, the software puts the lines in a slightly different order from the original if the size has been changed. Forget trying to match back seamlessly to the original video if the scan lines are in slightly different places. You will also experience a loss of resolution because some scan lines are doubled to make up for missing ones. If you are exporting to re-import, always choose the native frame size and do not resize the image at all in the graphics program.

Exporting Metadata

An important trend in interchange between systems is the ability to export with rich metadata. Avid was critical in the development of metadata exchange by creating OMFI and implementing it as OMF 1 and OMF 2. This allows users to retain more of the creative decisions when they move to a third-party program. The metadata creation is so important you can think of much of the editing process as metadata management. The interoperability with other devices in the distribution chain is critical to new applications for video. This is the reason Avid created the MetaSync feature and added a metadata track to the timeline. You will be able to program points in your sequence that line up with actor's individual dialog in scripts, multilanguage subtitling overlays, and interactive television applications not yet invented.

There are several steps to making the metadata exchange work correctly. First, the user must export from the Avid system using the preferred format. Up until now that has been OMF 2, but is currently AAF. The third party must be able to import the metadata format and then, most importantly, the program must know what to do with the data! It is no good importing a rich metadata format like AAF and then stripping out all the information except for cuts and dissolves just because you don't know what to do with the information! Be wary of products that claim to import OMFI and AAF but only reduce it to an EDL once inside the program. You might as well just make an EDL!

Some manufacturers incorrectly made the assumption that OMF 2 was a closed Avid format. This left the door open for products like Automatic Duck's Pro Import®. This product takes the OMF 2 sequence metadata and opens it inside Adobe After Effects 5.0 and later versions. Common effects are transferred to After Effects like picture-in-picture, matte keys, speed changes, and collapsed layers. The user can link the metadata to the media already on the drives or you can export the OMFI metadata with the media embedded. You can quickly use the advanced compositing functions of After Effects and then render as a QuickTime movie with alpha channel using the Avid Codec.

AAF has been codified by the AAF Organization (www.aaf.org), which is made up of major manufacturers and broadcasters. It is largely based on OMF 2, but all aspects are controlled by this independent body to make sure that proprietary information is handled correctly. A common misconception is that once everybody starts exporting and importing AAF, there will be transparency between applications. The AAF specification allows for the use of hidden or "opaque" data that benefit certain manufacturers that choose to share the ability to unencrypt this proprietary metadata. And even if the manufacturer chooses to make certain information public to all, there will be varying degrees of success as third parties try to incorporate the data correctly into their programs.

Avid has implemented the export of AAF already for several releases. This will enable Avid to share data better with Avid DS and Digidesign Pro Tools® as well as any third party that correctly implements the standard. Currently, there is an extremely high degree of information shared with Avid DS that can be used for conforming high-definition masters from Avid offlines. This interchange saves approximately ten minutes, conforming each complex effect that was created in offline compared to recreating the effect by eye. Multiply this time savings by the typical number of effects in a one-hour primetime program and you can see that using AAF between Avid applications can save hours per show during the high-definition online.

There are several choices when choosing to export metadata, described next.

Link to (Don't Export) Current Media

This allows any program that opens the metadata to know exactly where all the video and audio media are on the hard drives. It will automatically link without having to copy the media to a new location. This can be valuable when using a third-party program on the same system or working on a very fast network like Unity MediaNetwork. However, if you are working with uncompressed images, they are too big to play back over a standard gigabit ethernet network so don't try to link to current media.

Copy All Media

This option assumes you need to move the media to another system. You may want to copy the media files to a slower large drive just for transportation and then copy them again to fast drives once you get to your destination. If you are not careful, you may break the links between the media and the composition, so reduce the number of times you need to copy the media before you open the metadata in the destination application. If you break the links by accident you can use the Relink command. This choice is excellent if you want all of the original media of all the master clips used in the sequence.

Consolidate Media

By consolidating first you reduce the amount of media that must be copied and moved. You will reduce the length of the original master clips to a length only long enough to play the sequence and some user-defined handles. If drive space is low, copy time is critical; or, you are moving media over a slower network, so you will want this choice. You will not have the freedom to recreate the project from scratch because you will not have all the original media, but that may not be important at a late stage in the project.

Importing and Exporting Motion Video

The next step for import and export is using moving video. Usually, people want to import animations created by a three-dimensional (3D) animation program or an effect sequence rendered from a compositing program. There is also the demand to export for Web pages, CD-Rom, or for material to be used in a compositing program like After Effects. All the already-mentioned procedures apply, including resizing and safe colors, but with slight differences. Also, you must decide between two choices for formats: movies or sequences of stills.

Let's look at importing video first. The choice of whether to render as QuickTime, AVI, or PICT/TIFF sequences depends on several factors. The first and most important is: What format does your third-party application have as an export choice? Usually less-complex programs on Macintosh have only one choice—QuickTime. This does indeed make things simpler.

QuickTime

QuickTime is a format that serves many masters. Its primary purpose is as a distribution format, but it can also be used successfully as an editing and intermediate format. If you are exporting out of the Avid editing system for distribution via QuickTime, I recommend you export as a QuickTime reference movie and use a program such as the bundled Sorenson Squeeze® to process your video. Not only will it provide a high-quality compression, it can be configured via templates so you always get predictable results.

A QuickTime reference movie is a series of pointers, a small amount of metadata, about where the original media files are on the drives. When you export a reference movie you send this small amount of metadata to another computer on the network or to another program on your system. The other computer can use it for compression using a program like Sorenson Squeeze, or a hardware-based compression system like Anystream or Telestream. The compression system loads the QuickTime reference movie like it was a real QuickTime movie and the compression software looks for the media in their original location. It then grabs the frames and compresses the media without slowing down or disturbing the main editing system. This now becomes a very fast background task to create a movie for the Web or for a DVD.

In many cases, though, you'll use QuickTime as a transfer mechanism to send video or animated graphics between workstations, departments, or facilities. In this case, a regular embedded QuickTime movie is required as they won't have access to your media on their unconnected machine.

The Avid Codec

The key to getting a QuickTime movie rendered at high quality that imports quickly into the Avid is to use the Avid codec. A codec (compressor/decompressor) is a system extension that any third-party program can access because it is in a central location on the host computer. On the Macintosh OS 9 and earlier it is in the System folder along with other extensions, and in OS X the codecs go in System:Library:QuickTime, where System is the name of your boot drive. On Windows XP it is installed automatically in the C:\Windows\system32 folder. The Avid Codec is a .qtx file; it allows other programs to compress and decompress a rendered QuickTime movie using the Avid media file format. This means that even though you are rendering a QuickTime movie, you really are creating an Avid media file that is inside the QuickTime format. Technically, this is called *encapsulating* or *wrapping*. To work best, use the native frame size and do not mix resolutions in the exported sequence. You can change the resolution of the Avid Codec–based QuickTime movie during import if you must, but it will slow down the import significantly. This is especially important if you have only one resolution of an animation and you must use it for offline and online. With the Adrenaline and software only–based Avid editing systems you can mix and match resolutions in the same sequence, so create the animation at uncompressed quality and keep it that way when you import.

Make sure that you load the Avid Codec into any Macintosh or Windows system you are using for graphics or animation; you will be given Avid resolutions when you choose the quality of your rendered QuickTime movie. Send the Codec to your graphics people or to anyone who is subcontracting graphics for you. And by all means, make sure it is the most recent version! You can get that information by calling Avid customer support or by downloading from the Avid website (www.avid.com); be aware that the version of the Codec may change on a different schedule from the software itself. The reward of using the Codec and native frame size is the almost real-time import speed as you come back to the Avid system.

Size

Size of the frame is still a consideration for importing QuickTime, for all the same reasons. Refer back to the chart for proper frame sizes for the different formats. With a still image, maintaining the correct frame size is important because of the scan line relationship, but there is now another consideration—import speed. If you render at anything other than the native frame size, you have to double lines to make up for the difference or throw away resolution and waste rendering time. You are also wasting import time since Avid must resize each frame on the fly, which adds a considerable, unacceptable amount of extra time to the process.

The other choices involved, when rendering to import to Avid, have to do with field order and field rendering. To get the absolute best-quality rendered movie, choose field rendering if your application makes it available. Field rendering takes longer to render, but it is worth it if you are working with complicated video and lots of detail and you want this to be a finished product. This is because you want the movement of your animations to be as smooth as possible, and if you render using only frames, then you are giving up half the motion resolution. With field rendering you get all 50 or 60 fields available to you rather than 25 or 30 frames. The extra fields smooth out motion of moving objects by giving you more discrete images within the same amount of time.

If you are working in a 24p or 25p project then you are working in progressive frames so a field-rendered animation won't help much. In all other projects, however, you will gain significant quality improvements by taking the extra time to field render.

Importing with an Alpha Channel

With current versions of Avid editing software, you are able to import QuickTime movies with an alpha channel attached. Many Avid QuickTime codecs, including the DNxHD codecs, support alpha channels, enabling you to fast import these files into your system.

Using OMFI for ProTools

When moving a project with media to Pro Tools or AudioVision® there are few elements that are critical:

- All sample rates must be the same. You cannot mix 44.1 kHz and 48 kHz in the same sequence. If you have mixed sample rates you should create a copy of the sequence and convert the sample rate. Select all the sequences and choose Change Sample Rate under the Clip or Bin menu. If you are consolidating the media during the export you will have the option to convert the sample rate in the export dialog.

- Macintosh systems are not compatible with WAV files. If you have WAV files and are moving to a Macintosh or an older Pro Tools or AudioVision you may have to convert the files to AIFF-C. You will have to do this while embedding the media into the composition. When you choose OMF 2 export and embed you will be given the choice to convert the file to AIFF-C. You must embed the media when you export OMF or AAF in order to convert it from WAV to AIFF-C.

- Macintosh systems cannot mount drives striped together from a Windows system. A Macintosh system may have difficulty playing from any Windows-formatted drive. You may use a Windows-formatted drive for transport, but the media will be copied to faster HFS or HFS-formatted drives once the media arrive at the audio studio. You can mount HFS drives (even stripes) on a Windows system using third-party drive-mounting software like Mediafour's MacDrive®, but again there may be performance issues and the media might have to be copied. Find out the platform, format, and the sample rate preferred by your audio facility.

If you are moving the audio to a Digidesign ProTools session, there are two methods to consider. The first method requires less drive space and is faster, but the second method allows more flexibility.

The first method is to hand over the drive with the audio media files on it and create an OMFI file that is composition only. You may want to consolidate your sequence before you do this if you want to put all the audio media on another drive for transport to the audio workstation. This allows you to keep working with the audio files you have. You might want to lock your audio tracks in the sequence so you don't accidentally change something while the mix is going on.

The second method is to make an OMFI audio-only file. Consider this method if you are going to use software that does not recognize the Sound Designer II format that ProTools and Avid Macintosh editing systems use. This creates an intermediate file and converts the audio to another, more widely used audio format, AIFF, along with all the edit information. Once this very large file is moved over to the digital audio workstation, it can be converted back to a Sound Designer II file for use with earlier Macintosh versions of Pro Tools software. The intermediate OMFI file is opened in the OMF Tool that comes with Pro Tools and converted to a Pro Tools session. The original OMFI file can be deleted after the conversion.

The AvidLink for Pro Tools will allow you to choose either method. The AvidLink for AudioVision assumes you have the correct format audio file and saves only as an OMFI composition. You will need to copy the audio files to another drive manually or through a standard export dialog. Since AvidLink is a simplified workflow, if you have not created the audio files in the correct format you may have to use the more complicated method as outlined earlier.

To send the audio mix back to the editing software, you must "bounce" the audio tracks to a continuous audio track. This real-time process changes the audio file, which now has subframe edits, to a frame rate that can be used by the Avid. Alternately, you can output to a digital tape format, recapture it into Avid, and line it up to the beginning of the sequence. Having some sort of synchronizing beep tones with a countdown or flash frame simplifies this final sync.

Adobe After Effects

When preparing a composition in Adobe After Effects, you should always use certain settings when rendering for the highest quality:

- Always use 29.97 fps when exporting from Avid and rendering from After Effects in NTSC (not 30 fps). Of course, PAL is 25 fps and 24P is 24 fps.
- Graphics or other elements should be created at 720×540 (NTSC) or 768×576 (PAL). This is an accurate square-pixel representation of the television screen. Graphics are then resized in After Effects to the 720×486 (720×576) D1 pixel size.
- Whether one chooses to work in After Effects at 720×540, 720×486, or 648×486, it is vital that the final render takes place at 720×486. For PAL, the proper composition output is 720×576. This can be done by creating a composition in the final correct size, dropping your animation into it, and using Scale-to-Fit (Ctrl-Alt-F/Cmd-Opt-F).
- If you need to work at 16×9 you can use the widescreen selection when choosing the pixel aspect ratio under the new composition. This will keep the project the correct frame size for anamorphic standard definition. If you are moving material from Avid to After Effects and then back to Avid, using the widescreen pixel aspect will keep the material from being scaled twice.

- If field rendering in After Effects, choose upper field first when going to an ABVB system or a PAL Meridien system. If going to an NTSC Meridien system, it should be field rendered lower field first. All Xpress DV systems require lower field first.
- If working in 24p, do not field render! This is a progressive format and does not use fields.
- Using the Avid Codec is currently the best conduit for going back and forth between Avid and After Effects.

If rendering a graphic for compositing in an Avid, this is the best way to deal with the alpha channel:

- Always render a QuickTime movie with an embedded alpha channel because batch import allows for more control of the files once they're in the system.

When you render in After Effects:

- Select the composition in the render queue and choose "Add Output Module" from the Composition menu.
- One output module should be set up to save the RGB+, and set your color to Straight (Unmatted). Only DS uses premultiplied mattes in the Avid product line.

Importing and exporting video, audio, and graphics have many variations, formats, and choices. With this flexibility comes complexity, so any production company should find the processes that work best for it, simplify them as much as possible, and be aware that you have many tools at your disposal.

Conclusion

Learning to import and export graphics, animations, and metadata correctly from your Avid system means you position yourself as the hub of the creative process. You are in more control of the final result of any project and can confidently maintain optimal quality at every stage. You can collaborate better with graphic artists and animators as well as properly prepare your material for distribution on the Web or DVD. This makes you indispensable as both a technician and an artist.

8

INTRODUCTION TO EFFECTS

"Art cannot result from frivolous or superficial effects."

—Hans Hofmann

The effect capabilities of Avid editing systems are surprisingly deep and complex, and professional results can be achieved very quickly. One of the leaps forward in capabilities of the past few years is that the speed of computer processing units (CPUs) and hard drives along with greater PCIe bus bandwidth has made real-time effects more common. Many of the old restrictions of numbers of video streams and real-time effects have been shattered by the move to software- or host-based Avid systems. However, some of the reliability of the hardware-based systems is gone as well. Now the capabilities of your system rely on the configuration of off-the-shelf computer parts rather than custom-built Avid hardware. Numbers of streams and real-time effects may vary from system to system based on the host computer and not on Avid's hardware expertise. We will explore the capabilities and implications of this brave new world in this chapter.

The trick to maximizing your system's resources is to do creative work in real time and then render for superior quality. The last time I checked, time was still money, and the best way to use the time with a client present is to show them multiple versions and make changes quickly. The lines between cheaper/slower and expensive/faster are blurring, especially if the preview quality is good enough to make important decisions about the final version. Incorporating faster CPUs and hard drives means that effects done on Avid systems can compete with much more expensive workstations.

There is much to be covered to deal completely with effects—too much for this book—but some basics can get you past the beginner stage. There is nothing like the experience of a hands-on class, and

Avid offers several. This chapter can only hint at some of the techniques you will discover with enough time to experiment.

As you become more confident with effects, you will also become serious about nesting. Nesting is the feature that gives you more levels of video layering than you could ever practically use except for the densest of graphics sequences. Nesting gives you incredible power with an extra level of complexity. This chapter will discuss nesting after discussing the basics.

ACPL-Based Effects

Over the last few years the speed of the computer's bus has increased and CPUs and GPUs (graphics processing units) have increased exponentially. Computers are typically shipped with more than one CPU core or even multiple multicore CPUs. And GPUs now ship with dozens of processing units on a single card. All of this power adds up to some amazing real-time effects capabilities, and one of the significant changes made to Media Composer 3.0 was to replace the effects processing engine of the past using a new multithreaded engine known as ACPL (Avid Component Processing Library). ACPL effects are able to be processed on both CPU and GPU cores, fully exploiting the power in your computer. Naturally, these effects run best on the latest, fastest systems, but you'll also find improved performance on earlier-generation systems that have at least two CPU cores. (GPU effects processing requires the use of the latest-generation graphics cards.)

In addition, version 3.0 also includes multithreaded codec engines that allow the system to decompress video streams on one core and apply effects to them on another for an extremely efficient processing pipeline. When it comes time to render, this architecture provides for a blazingly fast render architecture that can render effects exponentially faster than previous generations of Media Composer.

Types of Effects

There are two kinds of effects: real time and non-real time. With modern systems, real-time effects are only constrained by the speed of your CPU/GPU architecture and drives. Depending on your system you may be able to get as many as ten or more real-time streams of standard-definition (SD) media and five or more real-time streams of high-definition (HD) media. When an effect cannot be played in real time, the Avid system will begin to skip frames, showing you as many frames as it can while maintaining audio sync. When this happens you'll see red dots on the time-code track, indicating which frames were skipped. This is extremely useful for quick previsualization of a complex effect. Don't worry about the skipped frames; when you render the effect you will get them all back!

Real-time effects have an orange dot over the effect icon when unrendered. One of the main advantages of a host-based (CPU/GPU) system is the ability to use the faster CPUs, GPUs, and large amounts of RAM to speed up any rendering. A faster PCI bus and an operating system that takes full advantage of all the bus speed available will also make a difference. In practice, all real-time effects become conditional on a modern Media Composer system. This is because real time is always determined by the capabilities of the computer and the context of the effect in the sequence.

A non-real-time effect is too complex to be dealt with so quickly and sports a blue dot once it is in the sequence. With the host-based systems, non-real-time effects tend to be some AVX plug-ins, and certain types of motion effects. The system will always try hard to play something, but the results may be unpredictable. Even fancy wipes that you shouldn't use anyway (a.k.a. "weasel wipes") will play a real-time preview. Clearly, if a real-time effect does the trick, it is preferable and your effect design should take this into consideration. You may want to substitute a real-time effect as a temporary replacement for the final non-real-time effect just to get the timing correct with a real-time preview. It is easy enough to replace one effect with another after all the multi-layered video and audio timings are perfect.

When layering effects on top of each other vertically, the material on the top track always has priority. This is not a true multi-channel digital effect device in the way most people think of standard DVEs. It is more like having many single-channel devices. Each video track can be considered a separate channel of effects and, with nesting, much more than that. It means that the separate video tracks do not interact with each other because they are each like separate sequences. This modularity allows quick exchanges of shots when it comes time to modify an effect. It also means that if moving objects are going to change their layering priority on the screen, then the lower object must be moved to a higher video track.

In this CPU-dependent world with so many shades of real time even the Digital Cut dialog has a choice called Video Effect Safe Mode. This checkbox is the last chance to make sure everything will play out to tape. As you would expect, this setting is a little conservative. If the system can figure out the minimum to render before being absolutely positive there are no dropped frames or

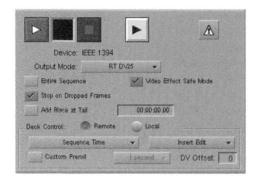

any other problem with the many layers of real time possible today, you should let the system take over. Think of the system as using a "manumatic" method of looking ahead and saving you time by doing the right thing before a digital cut.

Effect Design

Good effect design tries to achieve the most spectacular effects with the simplest use of layers. The fewer layers used, the fewer problems with trimming and rendering in a track-based effects model. Simpler design means most of the time you can modify faster because it is easier to figure out what is affecting what. If you can do something in fewer tracks, it looks better and renders faster.

Tree-based compositing is extremely powerful for creating graphic representations of the effect flow. This is the type of control offered by the DS Nitris system and some other third-party programs like Eyeon's Digital Fusion. You create branches by connecting effect nodes that could have mattes fed from one branch to another. Intermediate results (a traditional "work part" or submaster) can be connected to advanced controls without the restrictions of the Media Composer effect interface. The tree opens a whole new world for deep, complex effects that can be understood by this graphic signal flow. But all effects happen over time so there still needs to be a timeline and keyframe aspect that is tightly integrated with the tree. Consider signal flow, the order of effects, and the way they change over time when designing any composite.

Rendering

Rendering effects can be reduced significantly by using some basic strategies. In general, you render only tracks that are combined with non-real-time effects. With so much real-time capability these days, you should try to see if something plays without dropping frames before you consider rendering. Whenever you render an effect, you are rendering a composite of everything below. If you want to play the tracks below by themselves you can move the video track monitor down to lower tracks or stripping off the very top tracks to make multiple versions. Otherwise, you can leave these lower tracks unrendered. One simple method to rendering only a top track is to put a submaster effect on an empty track above the effect sequence. Put Add Edits in the empty track on either side of the area to be rendered and then drag a submaster effect between them. By rendering the submaster effect you are assured of rendering only the top track; if there are many effects sequences in a row this can be a time-saver.

What confuses people is that many times there is not just one track available to render as the top track. The beginning of the show may have a complicated layering section that has ten layers. Then most of the show may not go above track 3 and the end has five tracks. The best solution is ExpertRender™.

ExpertRender

ExpertRender is a feature designed to let the intelligence of the system solve your rendering problems for you. If the main problem with rendering is that people render too much, then the best solution would be to make sure everyone feels comfortable with a minimal style of rendering. The reason people render too much is that they don't really know what will play in real time and what won't. This is because many effects are conditional and depend on what else is going on in the sequence. Don't take the time to step through a long and complicated sequence effect by effect and still, perhaps, guess wrong. It is easier to mark in at the beginning of the sequence, out at the end, turn on all the video and audio tracks, and use Render In to Out. The expert part of the feature will leave as much real time as necessary and render only what is absolutely necessary.

If you don't use ExpertRender, then rendering in to out may be simpler, but there are some definite drawbacks. There are times when you may disagree with ExpertRender. Specifically, you may have plans for a certain section and you will be adding more effects to a higher track when you are done with the rendering. In this case, the system cannot read your mind to know what you will do next and can return only results based on the existing sequence. You can then choose to Modify the ExpertRender choices. By clicking on Modify in the ExpertRender dialog, ExpertRender leaves all of the chosen effects still highlighted in the sequence. You can Shift-select or Shift-deselect as you see fit and press the regular Render button when you are done. There is no need for a Render In to Out again because all the effects are already selected.

Another time you may disagree with ExpertRender is when you have dissolves between titles. This is relatively rare and really should be treated as an exception. In this case, the system will realize that the real-time dissolve cannot be played in real time, but because of the order that it must allocate resources, chooses the titles for rendering. Again, the user can override this situation easily, pick the shorter answer, and render the dissolve only.

Or you can dissolve titles using the Fade Effect button, which creates keyframes that do not need to be rendered. This limits you to fading up and down; if you want to dissolve between titles then the best method is still a dissolve.

The beauty of this automated analysis is that a vast majority of the time the choices are the shortest rendering answer. In reality, you will save so much time by letting ExpertRender do the job for you that even the occasional overrender is easily overlooked. How much time is wasted stepping through effects by hand? The guarantee that the system will be able to play the entire sequence after the ExpertRender process is, all by itself, money in the bank.

Partial Render

Partial Render is the ability for the system to render only part of an effect at a time and then come back later and pick up where it left off. This allows you to start a render at any time, even if you know you don't have enough time to finish rendering the entire effect. By pressing Ctrl/Command+. (period) you can escape from the render and keep or discard what has been rendered so far. This is especially useful if you have a series of slow, blue dot effects and you can start to render a little more anytime you take a break.

The system will create a new precompute for each partial render and tie them all together to play the final effect. You can see how much of an effect you need to render by changing the Render Range in the timeline view to show Partial. This is the most useful setting and should be left on most of the time since it will show you only what is left of an effect that has started rendering, but not quite finished. This is the default setting on current systems. If you change the view to All, then it will show you all effects in the sequence that are not rendered. Although this can be useful under some conditions, it can be confusing if you are relying on ExpertRender to figure out what needs to be rendered. The visual feedback in the timeline of the red or partial red line across the top of the clip with the effects can clash with the information from dupe detection. I would strongly suggest that Render Range display and Dupe Detection not be turned on at the same time. If these two functions are important to you, they can be made part of a workspace and changed with a single keystroke.

The only other drawback to Partial Render is that with long-term complicated effects projects you will be generating more precomputes. If you have been cleaning up precomputes once a week to keep the system operating without the high, unnecessary overhead of too many small files, you may want to do it more often. Most of the time, however, you won't even be aware that Partial Render is at work. The best features, many times, are the ones that make you more productive without attracting attention. Long after the sizzle of the product demo is over, you will be making your deadlines with projects you are proud of, and you won't really care why!

Keyframes

Almost all effects can be manipulated by keyframes. The only exceptions are the color corrections and a few other segment effects like flip and flop. Keyframes are the method to change an effect over time, and you always need at least two keyframes if you want the parameters to change. At the first keyframe, certain values about position, shape, or color are entered, and the settings change to match the values on the next keyframe. The change, if there is any, is smoothly interpolated between the keyframes. Keyframes can be added in Effects mode on-the-fly while playing and pressing the Keyframe key (the ' [apostrophe] key is the default on Symphony and Media Composer). Keyframes can be copied and pasted, dragged by holding down the Alt/Option key, and can be moved with the trim keys. If you want the effect to just hang on the screen, with no motion, then you can copy and paste the same settings between two keyframes or highlight both keyframes when changing parameters.

Much has improved with keyframes and some of the old advice no longer applies. Avid significantly improved the keyframe model for certain effects and will, over time, migrate that functionality to all the effects. Now the user can choose to work in the "classic" keyframe mode or promote the effect to the new keyframe model. You now have a keyframe per parameter—the ability to change each parameter with its own timeline and set of keyframes. Using the new model, you need only one keyframe if you want the parameter to change from the default but remain the same throughout the effect. As we will see in the next section, you also have a wide range of choices as to how the motion between the keyframes is interpolated and how trimming affects the timing.

Advanced Keyframe Model

Advanced keyframes add added power and complexity while preserving much of the old keyframe methods. You can choose to promote several effects to the advanced keyframe model as

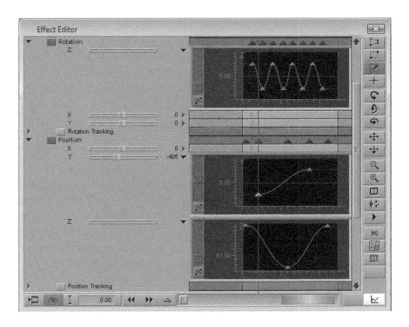

desired once you open the Effects editor in these later versions. Of course, you can keep the effect just the way it came to you from an older offline machine. But if you need to add some more sparkle to a project you can do much more while staying in the Avid program.

Look at the bottom of the effect in the Effects editor for this pink multi-timeline icon. By clicking on this all of your parameters are preserved (except Acceleration, which is replaced by something more powerful) and the keyframes are moved into a "timeline per parameter" effects interface. You can now add keyframes only to specific parameters and can add different kinds of motion interpolation that are much more sophisticated than simple ease-in/ease-out.

Creating, Deleting, Copying, and Moving Keyframes

When you move away from the concept of a single keyframe affecting all the parameters in an effect at the same time, you open some interesting possibilities that are also more complex. To maintain speed you need a series of choices for basic keyframe housekeeping. The first is a series of choices for adding exactly as many keyframes as you need.

Hold down the mouse over the Add Keyframe icon in the Effects editor and look at the choices. Let's look at why you would use each one.

Add to Active Parameter

This is the legacy choice of adding a keyframe to only the parameter that has been last activated by clicking on it. This choice is available if you have an active parameter chosen. It will add just one keyframe to this one parameter. If you have X and Y parameters of scale checked for Fixed Aspect, however, you will get keyframes on both parameters even with this choice.

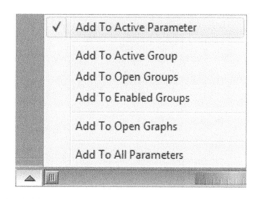

Add to Active Group

This will add a keyframe to an entire group of parameters like Position. If Position is Active (selected and highlighted pink) then a keyframe will be added to the X, Y, and Z parameters simultaneously. This is quite handy if a specific effect needs all axes to line up precisely at the same time.

Add to Open Groups

Like the previous choice, this will add a keyframe to all parameters that are part of a group that has been opened (the small triangle has been spun down to display all the sliders).

Add to Enabled Groups

When you promote a two-dimensional (2D) effect to a 3D Warp effect you get the enable buttons. These buttons allow you to adjust a parameter and then just disable it without resetting it to the defaults. This is a quick way to see multiple versions of the same effect since all those disabled parameters are still embedded in the effect, ready to be turned on to see the alternate version later. If several parameters are enabled you can choose to add keyframes to all of them with this choice.

Add to Open Graphs

This is a different way of thinking about convenience. You may have many parameters enabled in a complex effect, but to save screen real estate you have only the critical parameters showing the full keyframe graph. Rather than spend time scrolling up and down to make sure all unwanted parameters are disabled or closed, you can focus only on the open graphs where you are doing all the work.

Add to All Parameters

Not so sure about this advanced keyframe nonsense? You might want to go back to adding a single keyframe for all parameters and worry about which ones to tweak later.

Here are two sneaky shortcuts that we designed to make it even easier to use the power of keyframes per parameter. If you

right-click on the area of each parameter where the name of the parameter shows up in the timeline, the name of the parameter will change to the choice Apply to Group. This is a quick way to apply some change to the entire group without changing a single default setting. This can be applied to the entire effect by right-clicking on the very top of the parameter timelines, where the name of the effect is shown. This will change to Apply to All.

Deleting Keyframes

You can activate a keyframe and press the Delete button. You can also Alt/Option+click on the Add Keyframe icon to delete any selected keyframes. You can also right-click on any keyframe and get a menu choice for delete. You can even Shift-click to activate many keyframes and delete them all at once through any of these methods. Which way do you remember being the fastest?

Changing Parameters over Time

With the new graphs representing parameters changing over time you need some modifier keys to control the direct manipulation of the keyframes. You can just click on any keyframe and drag it up and down to change the parameters, but what if you want to move the keyframe sideways to change the position in time? You have several choices. If you like the mouse you can hold down the Alt/Option key and while dragging you will have complete freedom to move anywhere on the graph. However, you may now need to constrain such movement so the parameter doesn't change, just the placement on the timeline. In this case, you would hold down the Shift and the Alt/Option keys and now move only sideways. If the parameter graph is closed (click on the small left triangle to close a parameter graph) then the motion of the keyframe automatically will be constrained to time changes only. You only need to use the Alt/Option key. Finally, you can use the trim buttons that are mapped to the keyboard for a very accurate nudge. You can push all active keyframes one frame or ten frames depending on the trim key you use.

What is really interesting about this new method of displaying keyframes over time is that you can have keyframes before or after the effect itself. In other words, you can add parameters that will begin before the effect starts to play. This allows you to adjust timing with trimming of the clip with the effect. It is also a great way to have an effect match another effect by starting in the same place or same time and then syncing up later when both effects are later visible (landing at the same time or bumping against each other, for instance).

Aligning and Slipping Keyframes

Clearly there can be many more keyframes in each effect than ever before by using the advanced keyframe model. Chances are that you will need many of those keyframes to start and end at the same time. You need to align keyframes from different parameters so that they have a common point. This might be the beginning, end, or somewhere critical in the middle (like on the drum beat and cymbal crash). This is what the functions Align and Slip are for.

Imagine that you have added a keyframe to the Position X parameter that needs to match the Scale X parameter. You need the motion to stop at the same time the resize begins. Unfortunately, you have already created all the keyframes and realize that the effect is timed slightly wrong only after playing it back once.

The position keyframe is in the right place for the timing so it becomes the reference keyframe. Click on the reference keyframe to move the blue position bar to that location. If you have "Set Position to Keyframe" unchecked in the Effects editor setting then you will have to drag the blue bar to the reference location. Align always uses the blue bar as the point in time to match up. Then right-click in the Effects editor in the Resize parameter area and choose Align. The highlighted pink keyframes will move to the new position.

If you want to align more than one parameter at a time (like the X, Y, and Z parameters of Position) then you can use the sneaky shortcut of right-clicking on the name of the Parameter in the timeline above the graph or on the name of the effect at the very top of the timeline graph and choose Apply to All. This Apply to All is so powerful that you may end up affecting too many keyframes. Make sure that all the other parameters besides the ones you want to change have their graphs closed and no keyframes are highlighted pink. If you accidentally moved too many keyframes then undo the Align, go to those parameters, close the graphs, and Ctrl/Option-click on the pink keyframes to turn them gray. Then do the Align process again.

Slip is just like Align except that all the keyframes to the end of the effect are affected too. When the active keyframe moves to the position of the blue bar in the timeline, all the other keyframes in the effect stay in the correct relationship and shift the same amount. This makes sure that you don't change the timing of the rest of the effect when you line up one keyframe.

Trimming Effects and Keyframes

With the previous keyframe model you had only one choice of behavior when trimming a clip with an effect to make it longer or shorter. The keyframes followed along to make the effect slower or faster. This could be an amazing time-saver since it meant that the timing of the effect automatically followed the length of the clip, but it was also limiting. Sometimes you wanted the effect to just stay at the end of its trajectory, landing in just the right place on the screen at just the right time, but you need an extra beat or breath to absorb it before the cut. In this case, you don't want the timing of the trajectory or the moment of the effect landing to change when you make the shot just a little longer. This is why Avid created both elastic and fixed keyframes.

In a standard effect you can make any keyframe either elastic or fixed. You highlight the keyframe and then right-click to get the menu of choices. If you make a keyframe elastic then it behaves like it always did and changes timing with the length of the clip. If you choose a fixed keyframe then it will stay put in the relative timing of the effect no matter how you trim the shot. Now that the keyframe is fixed you need to determine what happens to the effect with the extra material in the shot. Does the effect stay absolutely still and hold in position? Or does it extrapolate and continue to move in the same direction of the trajectory? This all depends on the effect itself. If the effect has come to rest on the screen then either choice is fine, but if the effect was a slow move off the screen and you trim the clip longer and extrapolate, it will continue to move in the same direction a little longer.

Controlling Motion between Keyframes

By far the most powerful aspect of the advanced keyframe model is the ability to control the way the effect moves between the keyframes. There are now four different types of motion that improve upon the standard acceleration of 2D effects and Spline in the 3D effects. Let's explore each one and how they might be used.

Shelf

In many other programs Shelf would be referred to as Hold. When the user chooses Shelf it means that the effect stays in place until the time of the next keyframe and then it jumps instantly to the new position. This effect can be used to change a parameter when the object is hidden for a fraction of a second so that when it reappears on the screen it has changed. You don't

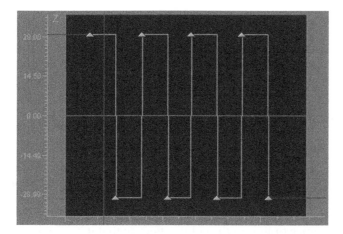

have to worry about the overshoot or undershoot of other motion types. You can also use it to bounce an object around the screen very quickly, but mostly it will be used to keep a parameter the same over time with a minimum of complexity.

Linear

This motion type is usually associated with very mechanical types of motion. The object moves between keyframes at a completely steady pace. There is no speedup or slowdown of the object, and this resembles the kind of unreal motion that only a robot could emulate. This type of movement is used when you have many objects moving at the same time, but they start and stop at different times and still manage to sync up. If objects are speeding up and slowing down at different times, it is very difficult to get them to land at the same time or combine into a graphic effect simultaneously. This motion type is also used when

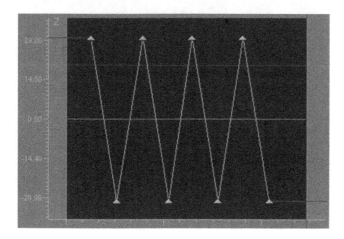

objects start and stop offscreen. If you are moving large letters across the screen so that they spell a word, you want them all to stay evenly spaced apart even though each letter is a different layer and 2D PIP (picture-in-picture). You also don't really want objects to appear to accelerate onto the screen and then decelerate as they exit. The illusion is that they are just passing by and not grinding to a halt somewhere just out of sight.

Spline

Those who have used the older 3D Warp effect are familiar with the Spline control. This was defined by animators to reproduce natural types of motion. Spline in this sense emulates the smoothest natural motion between multiple keyframes. There is enough intelligence built into the spline effect so that it can look over three or more keyframes to determine the smoothest path through all of them. As you move the keyframes, Spline automatically readjusts. At the simplest, Spline creates the ease in/ease out effect that is basic to DVE moves. It is simple and effective to create basic smoothness without complex handles.

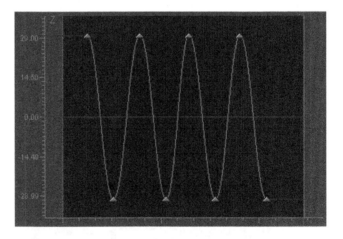

Bezier

Bezier begins as a Spline, but gives the user much more control—in some cases, perhaps a bit too much control, as it can create unexpected results from complex handles. There are handles on each keyframe that can be adjusted to control the amount of ease in/ease out, the speed of the velocity change, and make the effect behave differently on the other side of the keyframe. This is because a Bezier handle can be adjusted three ways: Symmetric, Asymmetric, and Independent. You can cycle between the three types by holding down the Alt/Option key when adjusting the handle. The effect defaults to Symmetric. Hold down the Alt/Option key, and click the handle to change the mode to Asymmetric. You can adjust the handle freely in this mode without a modifier key. If you hold down the Alt/Option key again you will cycle to Independent.

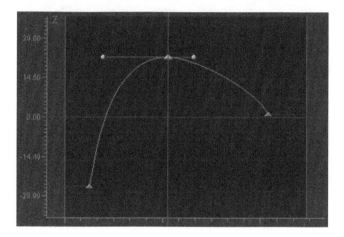

Symmetric

A Symmetric effect is created by pulling on both sides of the Bezier handle the same amount. This means that if the effect swoops and slows down to the keyframe, it will swoop away and speed up by the same amount.

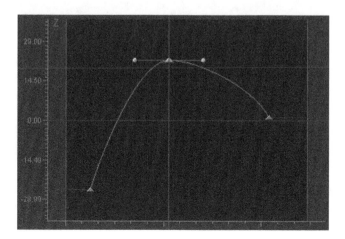

Asymmetric

You can adjust the handle to create a different speed curve before the keyframe and after the keyframe. By holding down the Alt/Option key while pulling on a handle on one side you can change the speed at two different rates.

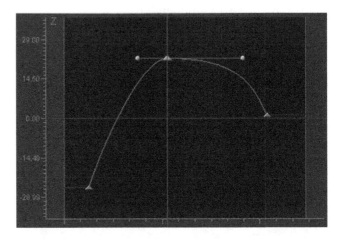

Independent

The Independent mode is sometimes referred to as breaking the cusp. This allows very different motion, sometimes quite extreme, on either side of the keyframe. Experiment with this mode for dramatic and unusual movement.

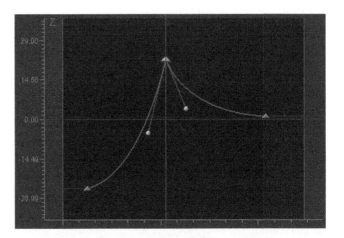

Copying and Pasting Keyframes

With versions prior to 3.0, you can copy and paste individual keyframes within a given parameter. But in version 3.0 you can copy and paste groups of keyframes within a given parameter and even copy and paste keyframes across multiple parameters simultaneously. Simply use the Shift key to select a range of keyframes then press Ctrl/Command+C to copy them, move to the desired position, and press Ctrl/Command+V to paste them. This technique was used to create the parameter loops seen in the previous illustrations.

You also have a new command at the top of the keyframe context menu that allows you to take a snapshot of all parameters at a given point in time and paste it into another. Simply park at the desired location, right-click in the keyframe region, and choose "Copy all Values at Position." The current value for all parameters in the effect will be copied and then pasted wherever you desire as a new keyframe. I use this technique to take a snapshot of a position in the effect that I want to return to or even land on at the end of the effect.

You can even copy and paste keyframes between effects, though not between different parameters. This means that you can copy and paste Position keyframes between effects, but cannot copy X Position keyframes and paste them into Y Position.

Removing Redundant Keyframes

Another new feature in the latest version is especially useful when conforming or troubleshooting effects created with basic keyframes. The Remove Redundant Effects command instructs the system to analyze the effect's keyframes and remove any keyframes that do not change the value of a parameter and are therefore unnecessary. Remember that adding a keyframe in basic keyframes adds a keyframe to every parameter, regardless of whether those parameters are being modified. These extra keyframes can get in the way when you are trying to either refine or troubleshoot an effect. I strongly recommend using this command whenever you promote a precreated effect to advanced keyframes. You can apply it to individual parameters, but I recommend right-clicking on the top of the keyframe region so that it analyzes and cleans up every parameter in the effect.

Effect Editor Settings

Now that we have looked at all the capabilities of the advanced keyframe model, we should take a close look at the settings that control the interface. Open the Effects editor setting in the Project window to see the new choices. These settings are a combination of controls for display and use of screen real estate along with performance enhancers for slower machines. The first three choices—indent rows, large text, and thumbwheels—make the text easier to read on high-resolution monitors, and the thumbwheels save

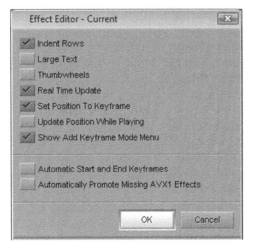

valuable horizontal space that is used up by the classic parameter sliders.

The next four choices are ways of turning off displays so that they don't try to update the user interface during the creative process. If you have a good idea of what you need to change in the parameters, then having the computer slow down to try to display those frames is not that valuable. The Set Position to Keyframe control, however, can be turned off if you are doing a lot of align and slipping of keyframes. This is because you are clicking on keyframes for the alignment and can work faster if the screen doesn't try to update a complex effect every time.

The Show Add Keyframe Mode menu is also a time-saver. You can set a default for the Add Keyframe button (again, mapped to the ' [apostrophe] key by default) or for the Add Keyframe icon in the Effects editor window. If you find that you need the flexibility of adding keyframes to open groups and some of the other more advanced choices, you will want the Show Add Keyframe Mode menu to be on. A single click on the Add Keyframe icon brings up a menu of choices. Click twice to take the checked choice or change the choice depending on how many keyframes you want to add. The Add Keyframe keystroke will follow the choice made in Mode menu.

The final choice in the Effects editor Window, Automatic Start and End Keyframes, gives you a choice between preserving classic Avid behavior and moving completely into the more advanced new methods. The old Avid keyframe model always had two keyframes and, although this was at times comforting and familiar to the experienced Avid editor, it didn't follow the model of other programs. If there is no change in the parameters over time then why add a keyframe at all? The first and second keyframes would need to be highlighted to make any change that applied to the whole effect and, more likely than not, you would forget to select one and end up with some unwanted animation.

As soon as you promote an effect to the advanced keyframe model you don't need a keyframe to change the default effect parameters. Just change the parameters without one. If you add a single keyframe you are not held back by needing another one at the end of the effect. This simplifies basic parameter changes and makes simple trimming situations even simpler. I would uncheck this choice and move completely into the advanced keyframe world.

Timewarps

The traditional "Source Side" motion effects have been replaced by a much more advanced time-warp control for creating motion effects in the context of the timeline. Timewarps are now applied through the Effect Palette like all other effects. However, you do not have to be in Effects mode to use it. This small fact becomes very important later as we look at the interrelationship between time-warps and trimming. The important thing about the new motion effects user interface and the time-warp effect is that you can use the advanced key-frame model to control speed changes. You also have a wide range of techniques for complete control over remapping time. Additionally, you have a wide range of new types of motion, which elevates this particular technique to an art form needing just the right touch.

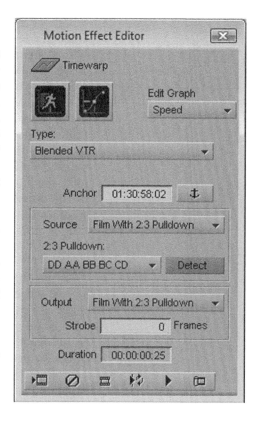

There are two types of control over the time-warp: speed and position. Most people will use speed since it makes the most sense for a wider set of circumstances, but both are very useful.

Speed

This pane allows you to add keyframes mapping speed to relative time. Although this sounds like a mind-bending concept out of Dr. Who, it is really straightforward. You add a keyframe at the time you want to change and move it up or down to determine the speed. Add another keyframe and now you have a ramped speed change. You can start the speed at 100 and then ramp it down to 0 to have a smooth transition to a freeze frame. You can add as many keyframes as you like to change the timing and control the smoothness with the wide range of motion interpolation choices discussed above in the "Advanced Keyframe" section.

You need an anchor frame for the speed change to really work. Fortunately, you will always have the first frame of the effect mapped as the default anchor frame. This is the frame that doesn't move when everything around it is changing. It is the frame that stays in exactly the same place as you change the key-frames around it. This is critical to having a motion effect that doesn't have the first frame change when you make it slower. You have picked the first frame, and you are making the frames in front or behind it speed up or slow down. Keeping the first frame exactly at the beginning of the effect gives you a reference point you can count on.

You can move the anchor frame if you want to get more sophisticated. This would be critical if you wanted a shot to start at a specific place and go backward. You don't want to put the first frame of this backward effect at the beginning of the cut! Edit the clip in to the sequence by making the outpoint of the source clip the first frame of the reverse motion. Go to the last frame of the clip in the sequence (the starting point for the backward motion), and click on the Set Anchor Frame button. Then change the speed to −100. All of the frames in the effect are visible in the timeline; they just go in the opposite order. Experiment with the anchor frame if you want a specific point to be matched to audio cues, and have all the frames around it update with speed change keyframes.

Position

Position is even a bit more mind bending. With Position you really are mapping a source timecode to a sequence timecode. This is an excellent choice if you have multiple points that have to hit exactly to a cue like a music sting or an explosion sound effect. You go to the exact frame of the source material and add a keyframe. Then move that keyframe so that it matches the chosen point in the sequence. Keep adding keyframes until all the points are mapped to the proper places.

Timewarp and Trim

Unlike other systems that allow you to keyframe motion effects, the Avid system is designed to keep you from destroying the rest of your sequence! Imagine that you have a full-length project with lots of audio and video edits. You add one motion effect in the middle and decide to make it just a little slower. You certainly don't want to knock the rest of your sequence out of sync! But other systems seem not to care that whenever you change the speed of the clip you are changing the length in the timeline, too.

Avid designed a different solution. Here, when you change the speed of an effect you do not change the length of the clip in the sequence. You need to go to Trim mode and make the adjustment in a controlled, predictable way using the best trim model in the business. Don't leave the result to chance when you change speeds, especially when you have multiple keyframes and you are spending lots of time experimenting. The ability to adjust the speed of a clip while you are in Trim mode works quite well as a basic technique.

Formats

You can change formats of the image during the speed change. This is very important when working with 24-fps film in an NTSC 29.97 project. Since all 24-fps film has a 3:2 pulldown when transferred to 29.97 videotape, these extra frames become very visible when the speed gets slower. Generally our eyes compensate, but as soon as you extend the amount of time the pulldown frames are on the screen the motion looks very jerky. The best approach is to remove the pulldown before creating the motion effect. Go to the Formats section and choose "Film with 3:2 pulldown." The system will then try to detect the cadence of the pulldown since the beginning of the edit is probably not at the A frame or beginning of the pulldown cadence. The system is very good at detecting where the duplicated frames are and removing them. However, if it makes a mistake or is fooled by an animation that starts on repeated frames, then you can override the cadence detection and try it on your own. Choose the 3:2 pulldown to reinsert the pulldown after the effect is created.

If you have shot progressive frames on video you can actually add in 3:2 pulldown for a film look. Choose "Progressive" as the input and "Film with 3:2 pulldown" as the output. The system will assume the first frame of the clip is the A frame and add in the duplicated frames. You could even choose "Interlaced."

Motion Effect Types

With so many different types of motion effects available we need a quick overview of what each is doing and when you would use it. Many of these effects are real time within certain restrictions on the host-based systems.

Duplicated Fields

This type drops out the second field of each frame, which significantly softens the picture. However, this type is excellent for experimenting with motion since rendering Duplicated Fields is much faster than any other type. To get a really fast render you should experiment with the 3:2 pulldown still in the image. Once you have the motion close to the way you want, remove the 3:2 pulldown with the Format button and change to a higher-quality motion type. Think of this as the draft mode for motion effects. You may receive offline sequences from older systems with this choice selected by the offline editor. Chapter 10 goes into detail about how to accommodate this choice when rendering in online.

Both Fields

Although this method preserves both fields of every frame, it almost always looks choppy when used on moving material. It is excellent for preserving sharpness of a still image that you must extend to fill a sound bite.

Interpolated Fields

Although this creates smoother motion than the previous two choices, this method is also slow to render and somewhat soft. The system is making mathematical calculations as to which actual fields to combine together into a single frame for smoothest motion. Unfortunately, you may get field 1 from frame 1 and field 1 from frame 3 combined into the same resultant frame. Since the frame never actually gets the information from field 2, it will always look slightly soft. This combining of fields is rather random depending on the speeds you have chosen, but to keep the image from bouncing back and forth from soft to sharp, all of the frames are slightly soft.

VTR-Style

This method reproduces the type of motion effect a VTR (video tape recorder) would create when playing back each field. This method is sharper than interpolated and smoother than both fields, but isn't doing any fancy math to compensate for jitter. You will see a little horizontal movement as you go between fields. Since so many people are used to seeing this from VTR playback, it may not even be noticeable under medium speeds. At very slow speeds, however, you will see a difference, and interpolated may be a better choice.

Blended Interpolated and Blended VTR

This modification of the previous styles tends to smooth out the motion even more. It averages the frames and performs somewhat of a dissolve between frames. Although it preserves the best aspects of the original motion type, the blending occasionally can call attention to itself as a "look." But if the image is looking jerky, try this twist.

FluidMotion™

Someday all motion effects will be as smooth and sharp as FluidMotion. This computationally intensive breakthrough in motion effects actually makes new pixels from combining original source pixels. If you want to make the smoothest possible motion effect you need to recreate the look of a high-speed frame rate. Because usually you have only a limited number of frames from normal film and video production, you need to manufacture the in-between frames that will eliminate the jerkiness of normal slow motion. Blending and interpolating will get you only so far; then you need to predict pixels.

FluidMotion makes new in-between frames by looking at the real frames before and after and tries to predict where each new pixel should be. It can combine any number of frames together as long as it properly tracks the individual pixels and in what direction they are moving (their motion vectors). The problem arises when the system can't predict what the next frame will look like. This happens with extremely fast motion where the pixels from the real frames jump huge distances in a single field. The pixels change so radically from frame to frame that the system cannot track them (a similar problem arises in the tracker controls in Symphony and DS Nitris). It also happens with occlusion, or when part of the image is covered up by a foreground object that is not moving, like a tree. This is because the prediction from one frame to another is interrupted if important pixels just disappear for a few frames and then show up with no history. If a pixel pops out from behind a tree you can predict where it is going only by looking into the future. The basic algorithm of looking into the past as well as the future to make a weighted decision about predicted in-between position is disrupted. FluidMotion doesn't know about trees, only pixels moving through time and ones that don't. This also applies to objects entering and leaving the frames.

When the pixels go astray they appear to morph objects into each other. There are some fascinating controls to help correct for these pixel-prediction problems. The user draws around the problem area on a frame-by-frame basis and forces the area to have a certain vector or direction. This is done through a basic paint tool and an eyedropper.

Stop on the frame where the image is morphing incorrectly. Click the paintbrush icon next to the FluidMotion choice in the Timewarp interface. You are shown an analysis of the vectors in the image—what direction each object, really each collection of pixels, is supposed to be going. The direction, or vector, is represented by a color. The colors are mapped to the points of a standard vectorscope like directions of a compass. If an area is yellow then the FluidMotion effect is predicting that the pixels in that object are moving to the left. If this pixel prediction is incorrect, use one of the drawing tools to draw a selection around the yellow object. The selection will turn gray, showing that it has "zeroed out" the vector. While the rest of the image will have its motion predicted, this section will act more like a blended interpolated effect. This may solve the problem. Feather the edge of the selection and render. If it doesn't then you need to grab the eyedropper over the color-selection tool of Set Vector mode. With object selection still active, grab a color from somewhere else in the image that matches the correct vector. If the object should really be moving up in the picture, then grab the color of something moving in that direction. Up would be red on the vectorscope/compass. The selection will turn red and you can feather the edges a little to make sure it blends correctly. Render the effect and see if it does the right thing.

FluidMotion is an excellent choice to hide the fact that there is a motion effect at all. However, it does have its own look and can be used quite effectively to make something look unique.

Timewarp Freeze Frames

Though you can certainly create freeze frames using the standard Freeze Frame effect, one significant problem you'll eventually encounter is that you cannot make any changes to this effect, including changing the render method (e.g., from the default of Duplicated Field) or changing the frame frozen, without recreating the effect. The render method limitation can cause significant problems in an online conform. Though the Timewarp effect does not explicitly include a "freeze frame" option, you can easily create a Timewrap Freeze Frame effect that is much more flexible than the standard Freeze Frame effect.

To create a Timewarp Freeze Frame effect:

1. Park on the freeze frame in the timeline, and, if necessary, turn off all higher tracks.
2. Use Mark Clip to mark the existing freeze frame in the timeline.
3. With the timeline active, use Match Frame to load the motion effect in the Source monitor.
4. Use Match Frame again to load the freeze frame's source clip.
5. Overwrite the source clip over the freeze frame. It is possible that sufficient duration does not exist within the original clip to fill the freeze frame duration, especially if the offline sequence was decomposed. In this instance, overwrite as much as is available. Then, after creating the Timewarp Freeze Frame effect, trim it out to the desired duration.
6. Apply a Timewarp effect to the clip edited into the sequence and enter Effect mode to open the Motion Effect Editor.
7. If it has not already been set as the render default, set the Timewarp render method to "Blended Interpolated." This rendering method is the best one for most freeze frames. If there is no intrafield motion, you may want to use "Both Fields" instead.
8. Open the Speed graph.
9. Set the speed of the active keyframe to 0. In addition to dragging the keyframe, you can use the leftmost field at the bottom of the Motion Effect Editor to quickly set a value of 0.

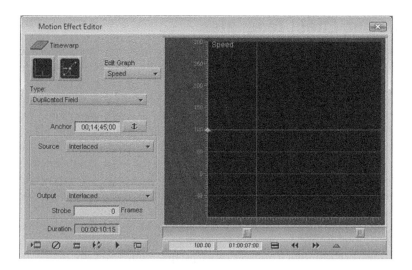

10. If necessary, trim the effect to the desired duration. If you have multiple freeze frames to remake, you should save the created effect to a bin so you can reapply it later.

Modifying a Timewarp Freeze Frame

Though we created the freeze frame using the first frame in the edited clip, we can actually use any frame that was edited into the sequence. To change the frame that is frozen in a Timewarp Freeze Frame effect:

1. Park on the Timewarp Freeze Frame clip and enter Effect mode to open the Motion Effect Editor. If your sequence contains multiple video tracks, make sure that the track containing the freeze frame you wish to modify is active. If it isn't, the Effect Editor will open instead of the Motion Effect Editor.
2. Open the Speed graph and set the speed of the active keyframe to 100 so you can access the other frames in the clip. Optionally, you could set it to a higher speed or even a negative speed. These rates are particularly useful if the desired frame is not within the portion of the clip that was edited into the sequence.
3. Park the position indicator on the frame you wish to freeze and add a keyframe.
4. Press the Anchor button to affix it to the new keyframe. The Anchor locks the source frame to the keyframe and ensures that it will always be displayed at that keyframe.
5. Delete the first keyframe in the effect as it is no longer required.
6. Select the new keyframe and set its speed to 0.

Saving Effect Templates

After all of this work making the effects just right, you can save them as effects templates so they can easily be applied over and over again. Any effect can be stored in a bin by clicking the effect icon in the upper left corner of the Effects editor, and dragging and dropping the effect icon in the bin. If a bin that holds an effect template is open, that effect will be available in the Effect Palette. The effect template can also be dragged back from the bin and applied to the timeline. When you apply the effect, it looks slightly different from the original effect unless the new clip you apply it to is exactly the same length as the original clip, or you have chosen fixed keyframes for all the parameters.

There is a sneaky trick very few people know for applying just part of a template. Using the Effects editor, open a specific parameter. You can drag and drop an effect template directly onto the single parameter. The open effect will take on only the parameter that came from the template, not the rest of the effect. This is very useful for matching drop shadows or border color and width. This also works with the color-correction mode if you just want to repeat a hue adjustment but nothing else.

If the Alt/Option key is held down when dragging the effect icon to a bin, then the effect template is saved with the video (segment effects only). This is an "effect with source" and can be edited into the sequence like a master clip. If you add an effect to a title, then save the effect template for the title always "with source," so you don't need to hold down a modifier key. If you want the title effect template to be just the keyframes alone so the effect can be applied to another title, hold down the Alt/Option key when saving it. Here's how to remember it:

- Alt/Option drag the effect template for effects gives you an effect and the source clip. Use this like a subclip with an effect attached.
- Alt/Option drag the title template on titles gives you the title's keyframes and no source. Apply this template to another title to get a similar title move.

Add Edits

If an effect cannot be manipulated with keyframes, it can be split into sections using the Add Edit button. By splitting one effect into multiple effects, each one can be manipulated separately and then recombined by a dissolve. This is most useful for a color correction that can be used to change a color over time with add edits and dissolves. If you have a camera that moves from exterior light to interior or from bright sun to shadow you can mix the two color corrections seamlessly. This is much easier to do with a dissolve than with complicated keyframes since it is essentially two complete setups rather than selective parameter changes. The system will play both color effects and dissolves in real time.

Creating add edits adds extra keyframes to an effect sequence. If you have an effect with two keyframes and you split it with an add edit, then you have two effects with two keyframes each. If the original effect had a smooth motion across the screen and now it has double the keyframes, you could have a problem with acceleration on a basic 2D effect. Acceleration is an effect parameter that smoothes the motion of an object's path across the screen with an ease-in/ease-out speed change. Adding an extra keyframe causes the effect to slow to a stop at the new point and pause where there used to be a continuous motion. If your effect has an object in motion with acceleration, you should apply only an add edit at a point where a keyframe already exists.

Nesting

Nesting is the most complex, but the most powerful, characteristic of effects with the Avid editing systems. So far we have discussed building effects vertically, which, depending on your model system, may be limited by the number of tracks. Once you understand nesting, you can expand the amount of tracks dramatically. Nesting involves stepping into an effect and adding video tracks inside. More effects can be added inside the nest, and then you can step into those. It is a way of layering multiple effects on a single clip, but also much more. The only real limitations are how long you want to render and how much RAM you have.

There are two methods to view a nest. You can apply an effect and then use the two arrows at the bottom left of the timeline to step in or out (these buttons are also mappable to the keyboard). Once inside the nest you can no longer hear audio, but you can focus on that level alone and work on it like it is a separate sequence. Within that layer you can add as many new video tracks as your model allows. Red numbers on the timecode track in the timeline will indicate how many layers you are nested in later versions.

The other method to view nesting is used on all models. Using the segment arrow to double-click on a segment with an effect, the tracks in the timeline expand to see all the layers inside at the next level. Continuing to double-click layers inside the first effect reveals those tracks as well. Tracks can be patched and edited in this mode, and the audio can still be monitored. It is a little easier to understand all the effects going on because the display is more graphic.

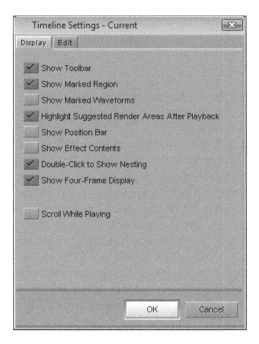

This mode of viewing nesting can frustrate people who open it up by accident and then are confused about what they are looking at. You can always close the expanded view by double-clicking on the original track again with the segment arrow or, in Media Composer or Symphony, Alt/Option+clicking on the down-nesting arrow. The view can be turned off in Media Composer and Symphony by unchecking the checkbox in the Timeline Settings called Double Click Shows Nesting. I recommend turning this feature off if you like to move very quickly and you have an older Mac.

Auto-Nesting

Nesting as just described implies a certain order of assembly. Apply the outer effect and then step inside. You must apply the PIP and then step in for the color effect. But real life doesn't always work this way. Many times the nest is a secondary thought, used well after the first effect is in place and rendered. In this case there is auto-nesting:

1. Select the clip with the segment arrow in the timeline.
2. Alt/Option-double-click an effect in the Effect Palette.

The second effect does not replace the first effect, but it covers it. This adds the layers from the outside instead of stepping into the effect and building them from the inside.

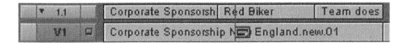

All these methods are for adding multiple effects to a single clip, but they are just as useful for adding one effect to multiple clips. If you want a color effect to cover an entire montage, then it is a waste of time and energy to put a separate effect on every clip. What happens when the effect must be changed? Now you need to change one, turn it into a template, and apply it to all the others. But there is a faster way that uses auto-nesting:

- Shift-select multiple clips in the sequence with the segment arrow.
- Alt/Option-double-click on the effect in the Effect Palette.
- The effect auto-nests as one effect that covers all the clips.
- Adjust the one effect and all the clips are changed.

If you want to change one of the clips and replace it with another shot, just step inside the effect and make the edit. You can also step inside the multiple-clip effect and add dissolves or other transition effects. You must render these inside effects, but you can leave the outside effect in real time to allow for future changes. Of course, if you render the outside effect, it will create a composite of everything inside.

The main drawback to this method is that you will have to step inside the nest to trim the clips. But if the trimming stage is long over and you are tweaking and finishing this is not such an issue. With the color-correction mode it is faster to save a correction as one of the four "buckets," so you may prefer to use this mode instead of nesting. By mapping the buckets to function keys you can move just as quickly through short sections. The Symphony's Program side color correction is even easier by using the "Use marks for segment correction Color Correction" user setting described in Chapter 12.

Viewing and Changing Nesting

Collapsing Effects

You can nest an entire effect sequence into one effect after it has been built using vertical layers. If keeping track of all the video layers becomes tedious, you can collapse them into a single layer. In order to nest effects, you must have an outside effect at the outermost level so collapsing places a submaster effect over all the layers and nests them inside. Select the area to be collapsed by marking in and out and highlighting the desired tracks. Then press the Collapse button and watch the animation.

There is no way to really uncollapse an effect segment. Here is the best method to work around it:

1. Step inside the collapsed effect.
2. Mark an in point and out point around the entire segment.
3. Turn on all the video tracks (except for V1 if it is empty).
4. Use the Copy to Clipboard button while inside the collapse.
5. Paste the clipboard contents into the Source window. In Media Composer or Symphony you can use the Alt/Option key when copying to the clipboard, and pasting to the Source window happens automatically. The layered segment can be used as a subsequence.
6. Cut the subsequence back over the top of the collapsed effect in the timeline or drag it to a bin.

This alternate method is actually even simpler:

1. Create some new video tracks—the same number as in the collapsed effect.
2. Expand the nest in the timeline by using the double-click with a segment arrow method to show all the tracks.
3. Drag the segments up to the empty tracks using the red selection arrow, and the Control (Windows) or Command (Macintosh) key to make sure they don't slip horizontally.

In Avid Media Composer 3.0 you can select all of the segments at once and move them up to the new tracks!

You could collapse all the tracks except the top track, like a title, and leave it in real time so you can continue to make changes without re-rendering. Rendering a nested effect is simple; just render the top, outside effect, and the submaster. This leaves all the effects inside unrendered, but it is sufficient to play as long as you are monitoring that outside effect.

You don't need to leave the top effect of a collapse as a submaster. You can replace the submaster with another segment effect by just dragging and dropping from the Effect Palette. You can replace the submaster with a mask (to simulate 16:9) or a color effect. You can also step into a nest and render the top track inside the nest (or ExpertRender), thus leaving the outside effect in real time. You can keep tweaking the effect on the outside of the nest if the dissolves inside are rendered.

Collapse versus Video Mixdown

Although the collapse feature is excellent for simplifying complex effects sequences down to one video track, a collapse can still potentially become unrendered. If you are sure that an effect sequence will never need to be changed, match-framed back to an original source, or used for an EDL, then you can use video mixdown. Video mixdown (under the Special menu) takes any section of video between marked points, whether it has effects on it or not, and turns it into one new media file. This new media file has no timecode (which timecode would it use if you had 15 layers?) and breaks all links to the original media. This is why match-frame-to-original-source clips will no longer work and EDLs will no longer reflect the original source timecodes.

Video mixdown should be used only for finishing and for something that you will be using as a single unit over and over, like the graphics bed for an opening sequence you use every week. Once all the effects are rendered, a video mixdown is as fast as copying the media to another place on the drive, an insignificant amount of time. If the effects are not rendered before the mixdown, they will be rendered first as part of the video mixdown process so don't forget to count on the rendering time in your calculations. This workflow encourages rendering first and video mixdown later when everything is signed off.

A video mixdown will significantly improve the performance of Avid during long sequences with lots of effects. Instead of forcing the computer to "build the pipes" for complex effects with many short media files, it just needs to find one master clip. This means snappier reaction time when you press play. Always make a copy of the sequence before you overwrite a mixdown over your time-coded original sources, so when the client changes his or her mind, you will have a fallback sequence. Video mixdowns are very powerful and time-saving for a wide range of purposes, but don't use them for offline if you plan to recapture or make an EDL!

Chroma Keying

In a chroma key, you set up the shot in a studio to obtain precise control of the background, which consists of a flat, uniformly colored screen. The screen is usually blue or green. When you apply the Chroma Key effect to the clip, you select the screen color to key out, leaving only the foreground image. Because the effect removes the selected color from the image, the foreground subjects must not contain that color. The following list summarizes the requirements that give you the best results when creating a chroma key:

- The background should be flat, well lit, and of uniform color.
- The subject should be well lit and should not contain the color to key out.

- The video should be shot on a component tape format, such as Digital Betacam or Betacam. DV or HDV can be used, but only if the subject is well lit.
- If you're capturing from a dub, the dub should be a component or component digital dub of the camera master.
- When capturing, you should capture a serial digital or component signal.

There are four different chroma keys provided with the Avid Media Composer system and of them SpectraMatte™ is the most sophisticated—and capable—keyer. Located in the Key category, this is the highest-quality keyer available from Avid. The SpectraMatte keyer is designed to produce keys of material containing fine details, partial transparency (e.g., smoke and glass), and other hard-to-key foreground elements. It also includes sophisticated spill suppression and matte manipulation parameters. This keyer is available in Avid Media Composer Adrenaline HD 2 and later releases.

As it provides the best-looking keys, it is strongly recommended that the SpectraMatte keyer be used for the majority of keys. If 3D manipulation of a keyed element is required, you should use the 3D Warp Chroma Key instead. The RGBKeyer also has very good keys and, due to its color-correction capabilities, may be the best keyer in some situations. The basic Chroma Key effect is not recommended.

In addition, there are several third-party plug-in chroma keyers that can be added to the Avid system. These include Ultimatte AdvantEdge, Digital Film Tools's zMatte, and the Primatte Keyer. For additional information on third-party plug-in effects, visit www.avid.com. Plug-in keyers are typically non-real-time effects.

Using the SpectraMatte Keyer

Introduced in Avid Media Composer Adrenaline HD 2, the SpectraMatte keyer is the highest-quality keyer packaged with the Avid editing system. It is an advanced keyframe effect and standard keyframing is not available. Let's look at the parameters available for this effect and how they are used to generate a key.

Bypass

Use this parameter to toggle the effect on and off. Enable this parameter when you sample the key color in the foreground image.

Key Color

This parameter group is used to set the initial key color. Use the Eyedropper to sample the key color in the foreground image. If the color backing in the image contains a wide range of color

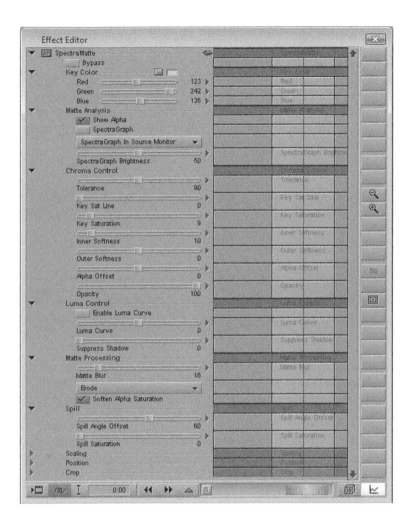

saturation or illumination, try sampling a color near the middle of the range of available tones.

You can use the RGB (red, green, and blue) parameters to tweak the sampled color. You can also use the Other Options button to display the operating system's Color Picker and use it to tweak the sampled color.

Matte Analysis

This parameter group is used to enable or disable the two matte analysis displays. These parameters do not affect the final key, but allow you to more easily tweak and troubleshoot the key. The available parameters are:

- *Show Alpha:* Displays the alpha channel generated by the key. Fully transparent areas are displayed as black and fully opaque areas are displayed as white. Partially transparent areas are displayed as gray. The intensity of the gray indicates the relative opacity of the area.

- *SpectraGraph:* Displays the SpectraGraph display for the key. The chroma values in the foreground image are displayed using a vectorscope plot. The range of chroma values that are being completely keyed out are overlaid in black. Partially keyed out areas are displayed as a gradient.

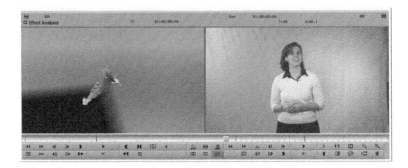

- *SpectraGraph Brightness:* Controls the brightness of the overlaid keyed region. Adjusting this parameter may make it easier to see the vectorscope plot.

Chroma Control

This is the primary parameter group used when creating and tweaking the key. The available parameters are:

- *Tolerance:* Controls the range of hues that are keyed. The greater the tolerance, the greater the range of hues keyed. The tolerance is centered around the chosen key color.
- *Key Sat Line:* Controls the saturation at which keying begins. This parameter is used to restore (or unkey) foreground regions that contain the key color, but at a low level of saturation. These regions are typically the result of the key color spilling onto the foreground. The Key Sat Line restores the low-saturation regions linearly.
- *Key Saturation:* Controls the saturation at which keying begins. This parameter is also used to restore (or unkey) foreground regions that contain the key color, but at a low level of saturation. These regions are typically the result of the key color spilling onto the foreground. Unlike the Key Sat Line parameter, Key Saturation restores the low-saturation regions by offsetting the keyed region. Though the Key Sat Line parameter is more typically used, Key Saturation has a specific interaction with the Spill parameters, which

sometimes makes it a better option. We will discuss this interaction later in the module.

- *Inner Softness:* Controls the falloff (from opaque to transparent) of the keyed region. This parameter is used to restore or remove regions that should be partially transparent in the key. At a value of 100, there is minimal falloff within the range defined by the Tolerance parameter. As the Inner Softness parameter is decreased in value, the colors at the edge of the keyed region are blended into the foreground instead of being completely keyed out. The default value for this parameter is 10.
  - Decreasing this parameter increases the opacity of partially transparent regions in the key.
  - Increasing this parameter increases the transparency of partially transparent regions in the key.
- *Outer Softness:* Controls the falloff (from opaque to transparent) of the pixels just beyond the keyed region. The parameter is used to adjust pixels that lie just beyond the boundaries of the key. The default value of 0 provides for a moderate amount of blending of the colors at and just beyond the edges of the key.
  - Increasing this parameter increases the transparency of partially transparent regions on the outside edge of the key.
  - Decreasing this parameter increases the opacity of partially transparent regions on the outside edge of the key.
- *Alpha Offset:* Moves the keyed area inward or outward along the axis of the key color. This parameter is not used as frequently as other parameters in this group, but can be useful when the color backing in the image contains a wide range of saturation values and you sampled from a color that had a high or low saturation value relative to the other areas of the backing.
- *Opacity:* Adjusts the overall opacity level of the foreground image. Use this parameter to fade the foreground image relative to the rest of the composite.

Luma Control

This parameter group contains postprocessing controls that can be used to further tweak a key. These parameters are typically only used to fine-tune particularly challenging keys. The available parameters are:

- *Enable Luma Curve:* Turns the adjustments made by the Luma Curve on and off.
- *Luma Curve:* Creates a luminance-based key postprocessor and is used when you need to manipulate the partially

transparent regions of the key and change their transparencies. The keyframe graph is used to define the transparency parameters that are to be manipulated. The Luma Curve postprocessor is not covered in this course.

- *Suppress Shadow:* Allows you to remove shadows cast by the foreground objects onto the background. Shadows typically cause the key color to have a low saturation value. Increasing this parameter increases the saturation values (that the keyer sees) for all pixels in the foreground. (The displayed chroma saturation is not affected.) This can make shadows easier to key out of the frame. However, it can also remove other partially transparent objects from the key and should be used sparingly.

Matte Processing

Your computer system may not be able to perform matte processing in real time. Using these parameters may require the effect to be rendered before it can be played at the Full Quality setting. If necessary, use the Draft Quality setting to preview the effect.

This parameter group is used to manipulate the alpha channel (or matte) generated by the keyer. It is typically used to soften the edges of the matte to make it composite more smoothly with the background. The available parameters are:

- *Matte Blur:* Sets the amount of blurring that is applied to the matte. Most effects require only a minimal amount of matte blurring.
- *Blur Menu:* Defines the type of blur defined. Three different blur types are supported:
  - *Blur:* Blurs to both the inside and outside of the matte edge.
  - *Erode:* Blurs only the inside edge of the matte. The original matte is mixed back in with the blur to ensure that the shape of the matte is maintained.
  - *Dilate:* Blurs only the outside edge of the matte. The original matte is mixed back in with the blur to ensure that the shape of the matte is maintained. This blur method tends to create a halo around the foreground object.
- *Soften Alpha Saturation:* Adjusts the blending of the two halves of the keying wedge. This parameter should be left on for most keys but disabling it might improve the edges of especially troublesome keys.

Spill

These parameters are used to remove chroma key background spill on the foreground subject. Spill is removed by extracting the key color from the pixels within the specified range. For example, if the background is blue, blue would be removed from any partially blue pixels in the foreground. The available parameters are:

- *Spill Saturation:* Only functions if the Key Saturation parameter in the Chroma Control group has been used. If you used Key Saturation you may have increased the visible spill on the foreground. This parameter restores the saturation values seen by the spill suppressor and, in conjunction with the Spill Angle Offset parameter, removes the excess spill.

• *Spill Angle Offset:* This is the primary spill suppression parameter. As you increase the value of this parameter, you increase the amount of pixels that are color corrected to remove spill.

DVE Controls

These parameter groups (Scaling, Position, Crop) provide a set of 2D DVE parameters that can be used to manipulate the image. These parameters function identically to those in the PIP effect.

If you need to access the 3D Warp DVE capabilities, you can nest one underneath the SpectraMatte. Be sure to enable the 3D Warp's Background parameter and set it to the backing color in the foreground image.

Configuring the SpectraMatte Keyer

In order to get consistently high-quality keys with the Spectra-Matte keyer, the following workflow is strongly recommended. Some keys will require variations from the recommended process, but following the process will ensure high-quality results for most keys.

Setting the Initial Key Values

1. Activate the Bypass parameter to disable the key.
2. Use the Eyedropper and the Color Preview box to sample the background key color. If the background contains a range of luminance and saturation values, try sampling from the center of the range of available tones. Alternatively, if you wish to preserve partially transparent foreground elements such as smoke, try sampling a color near the partially transparent elements you wish to keep.
3. Deactivate the Bypass parameter and enable the Spectra-Graph. Even though you may have carefully sampled the backing color, you should still check the SpectraGraph to help determine whether the selected key color is properly centered in hue in relation to the range of background tones. The SpectraGraph makes it easy to determine this.
4. Open the Key Color parameter group, and use the Red parameter to recenter the keyed region in the range of hues. For both blue-screen and green-screen keying the Red

It is best to view the SpectraGraph when you are making these initial changes as the adjustments make what appears to be a relatively subtle change to the key result at this stage. However, if you don't make these adjustments, the fine-tuning adjustments you need to make later will be much harder or even impossible to do.

parameter usually provides the widest range of adjustment. For certain blue screens, you may also need to adjust the Green parameter, and for certain green screens, you may also need to adjust the Blue parameter. Even if the sampled backing was highly saturated, moving the key color to the center of the saturation range will improve the key and make it easier to accurately key any partially transparent portions of the foreground.

5. Increase the Tolerance until the keyed region wedge includes all the key colors. Don't increase it too much or you run the risk of keying out colors that should remain keyed in.

Adjusting the Matte

Once you've set the proper key color and tolerance you're ready to adjust the matte generated by the key. In this phase you'll not only ensure that areas that should be fully opaque are opaque but you will also adjust the keying of the partially transparent regions in the foreground.

1. Disable the SpectraGraph parameter, and activate the Show Alpha parameter to display the alpha channel (or matte) generated by the key.

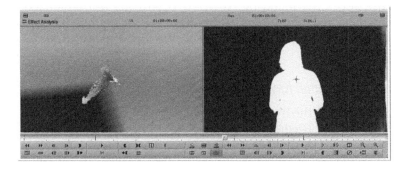

While adjusting you might want to toggle the alpha channel on and off to compare the result of the key to the alpha.

2. Examine the alpha channel, and compare it to the foreground image. Are regions that should be completely keyed out displayed as black and regions that should be completely keyed in displayed as white? What about the partially transparent regions? Are they displayed as gray or are they displayed as values of gray? In all likelihood the alpha needs adjustment. You will use the Key Sat Line (or Key Saturation), Inner Softness, and Outer Softness to adjust the matte until the image is correctly keyed. The exact adjustments of these three parameters will vary from key to key.

3. If necessary, increase the Key Sat Line parameter until the majority of the foreground is visible in the key. Once we've

restored the low-saturation regions, we will use Inner Softness and Outer Softness to ensure that the partially transparent regions are partially keyed. Most of these regions are currently completely keyed in instead of partially keyed.

4. Increase Outer Softness and decrease Inner Softness until you feel that the alpha channel correctly represents the partially transparent regions. If regions of your foreground image that should be completely keyed in are partially keyed instead, you should instead decrease Outer Softness and increase Inner Softness.

5. Disable the Show Alpha parameter and play through the key, checking all areas for keying errors. If you wish, you can also re-enable the Show Alpha parameter and play the matte.

6. If necessary, tweak the Key Sat Line, Inner Softness, and Outer Softness parameters until you are satisfied with the key at this point. Don't worry about any spill or hard edges on the matte. You'll correct those next.

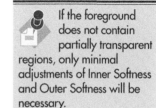

If the foreground does not contain partially transparent regions, only minimal adjustments of Inner Softness and Outer Softness will be necessary.

Suppressing the Foreground Spill

Depending on the color and shininess of the foreground objects, their proximity to the background when they were shot, and how they were lit, there may be a some color spill on the edges of your objects. Partially transparent objects will also exhibit spill (due their partially transparent nature). This spill can be removed using the Spill parameters.

The primary Spill parameter is Spill Angle Offset. This parameter defines a region surrounding the key region where the colors being keyed out are removed from the other colors in the image. For example, a partially transparent red object (such as a piece of cheesecloth) on a blue key background would contain a mix of red and blue. Increasing the Spill Angle Offset will remove the blue from the red/blue blend and replace it with a purer red. The larger the value of the Spill Angle Offset parameter, the more colors are replaced. Be careful with very large values of Spill Angle Offset with highly colored foreground images for it may affect the color saturation and balance in areas where spill is not a factor.

1. If necessary, disable the Show Alpha and SpectraGraph parameters.

2. Open the Spill parameters group.

3. (Optional) If you used the Key Saturation parameter, increase the Spill Saturation to an equivalent amount to reduce the exaggerated spill that may have been introduced by the Key Saturation parameter. This step is only required if you used the Key Saturation parameter; otherwise this parameter has no function.

4. Increase the Spill Angle Offset until the spill has been removed. Be careful not to use a heavy hand on complex foreground images. Be sure to look at the video monitor and external scopes to ensure that you are not overdriving the chroma of the foreground image.

5. Play through the effect to check the quality of the key and to ensure that the spill has been removed from the entire clip.

Adding a Blur to the Matte Edge

When you played through the key and/or alpha, you may notice an aliased (stair-stepped) edge on some areas of the composite. Depending on the source material, it may be necessary to add a slight blur to the edge of the matte. In most cases, the best blur method to use is Erode. This option blurs only on the inside edge of the matte. And, to ensure that the edge you defined is maintained, the original matte is mixed back in at a medium degree of transparency.

1. Open the Matte Processing parameter group and confirm that the Blur menu is set to "Erode." This option provides the best results for most keys.

2. Slightly increase the Matte Blur parameter. Most alpha channels only require a light touch on this parameter, especially if there are fine-edge details such as hair.

3. Render the effect so you can view the results of the key. Current computer systems cannot play SpectraMatte effects with matte processing applied in real time. Optionally you can switch to Draft Quality to play the key while tweaking the Matte Blur parameter. Ultimately, you should render the effect and view it in Full Quality mode to ensure that the key is satisfactory.

4. Play through the effect to check the final key. If necessary, you can return to any of the procedures to make adjustments and tweaks as required.

3D Effects

In Media Composer and Symphony, all 3D effects come from one effect, the 3D Warp. Two-dimensional PIPs, titles, and imported matte keys can all be "promoted" to 3D.

There is also corner pinning, which allows you to fit four corners of an object so that it matches the edges of another object. It is not quite morphing, but it allows you to put images inside TV monitors, picture frames, or the like.

There are so many things you can do with the 3D option that I helped write a one-day course just for that. Truly, this area calls for personal experimentation. Just taking some of the shapes and using them to warp images into interesting moving backgrounds requires parameters you must discover for yourself.

Paint and AniMatte®

If you consider editing to be an interframe process—working between frames—then painting on a frame is intraframe. There is a full palette of familiar choices for anyone who has used third-party painting programs. Brushes can be changed, and areas of the image can be blurred, color corrected, and generally affected like any standard paint program. Multiple layers of paint effects on the same frame are possible, but unlike the other, single-image paint programs, you can easily change the effects over time with keyframes.

You can also isolate parts of the frame, say the sky, and draw a matte shape around it to make it a deeper blue. The ability to draw on an object and create control points to adjust curves with Bezier handles and move the edge over time means that almost any part of an image can be manipulated separately from the whole. By creating points that change over time, any smooth, even motion in a shot can easily be followed with just a few keyframes. If the motion is jerky or unpredictable, you need more keyframes to adjust the control points. If the motion is truly difficult to follow or you are doing dozens of motion-following effects, you may want to use the tracking feature in Symphony that allows you to automatically track specific pixels over time. Since this is a complex and time-consuming task, occasionally you may want to set up a separate graphics station for tracking and rotoscoping. This way you can split the tasks among multiple people if time gets tight. You can also set up your graphics station to render while you continue to edit on the Avid system. Adobe After Effects®, Autodesk Combustion®, or Eyeon's Fusion® are good choices for this kind of work. DS Nitris is the best combination of graphics, paint, and 3D effects into a timeline-based editing program in the Avid product line. Paint and AniMatte are fine for shorter jobs, but if you need to do lots of this type of work you may want to consider a DS Nitris finishing suite.

AVX

An expanded range of choices is available with the AVX® (Avid Visual eXchange) plug-ins standard. This is an interchange format so that other third-party effects companies can modify their existing product line easily for use as an Avid plug-in. Make sure that the version of AVX effects you are using is also compatible with the Avid DS system if you want to take your sequence to the next step for high-quality, high-definition finishing. As computer processors get faster, there will be more you can do in real time with AVX effects.

Titles

Always try to work with titles at an uncompressed resolution in standard definition. Aliased edges and blockiness are eliminated when titles are uncompressed. If you are working on a Meridien system you have the ability to run the titles through the DSK. This is a special section of the hardware dedicated to being able to play uncompressed titles and graphics no matter what the resolution of the clip below. You also can run an uncompressed title on V2 and have unrendered real-time effects on V1 because you are not running three streams of video, just two streams and a PICT file through the DSK. With the host-based systems, your ability to play uncompressed titles is restricted only by the speed of your computer. You can mix compression types as much as you like with these systems.

Since you can copy and paste from a word processing program, instantly apply a custom style, and create title rolls on most systems, it is easy to use Title Tool for large amounts of text. You can create a custom title template and map it to a function key. All styles are mapped automatically to the next unused function key and are enabled only when the Title Tool window is active. If you are creating titles with lots of font and size changes, you can highlight the text in Title Tool and press the function key to apply the premade style.

You can open a title and edit it straight from the bin if you want to use it as a template for new titles. Ctrl/Command-double-click on the title in the bin to open Title Tool. It will also give you the choice of promoting this title to Marquee on some recent PC-based Avid systems. After you make changes you can Save As to create a title that needs to be edited into the sequence.

Title Tool has its own Safe Color setting. If you are picking colors from a color picker or trying to match a color in an image, it may automatically dull or change the color to make it broadcast safe. If you really need to match the color and take your chances later then turn off the Safe Color choice under the Object menu.

Title Tool is trying to show you a title for position, spelling, composition, and other basic choices as quickly as possible. This is why it defaults to a lower-quality draft mode during creation. If you want to see the finished quality of a title for approval purposes before you are finished, press Ctrl+Shift+P/Command+Shift+P to turn on the Preview mode.

The final tip for any title is to always check the kerning or leading of any title before saving it. This is the proportional spacing between letters and between lines, respectively. Basic fonts almost never have correct kerning on all words. You will have to kern the letters together or apart to make them even and aesthetically pleasing. The window marked Kern in the toolbar applies what a typographer would call "tracking" to the entire text string. You can do this from the keyboard by using the arrow keys and the Alt/Option key. Use the arrow keys to navigate to the font pair and then hold down the Modifier key to make the change. This will please both your inner and outer art director.

Marquee Title Tool

For truly complex manipulation of type, you will want to work with a program that can use vector-based graphics. Marquee now ships as a second choice for more sophisticated titles and animations on many models. When you choose Title Tool you will be given the choice between the old Title Tool and Marquee. If you want the choice to always be one or the other and you want to eliminate this pop-up option, choose Persist, and your choice will be remembered. If you ever want to go back to being given a choice you can go to the Marquee Title setting in the Project window. There you can switch to the other choice or reverse the Persist choice. If you have been given titles that were created on the old Title Tool you can choose to promote them in Marquee and continue to add that extra level of polish and pizzazz. You can also continue to use a mix of old titles and Marquee titles in the same sequence. Once you have promoted an old title to the Marquee title you cannot go back; this may have implications if you want to send your finished sequence back to an older system for continuing offline work. There is a checkbox for saving a version of your original title before promoting that you should check "on." If you need to get to an older version of the title before you go back to a version that does not support Marquee, you can cut these titles back into the sequence.

Marquee, a true 3D type and graphics manipulation program, allows you to quickly create titles with textures, light sources, and extruded type. It will give you real-time rolls and fast render crawls on Meridien systems. You can manipulate each letter in a title on its own timeline and control all the movement with Bezier curves. A static title is quick to create and plays back in real time. An animated title will take longer to render, but it will preview in real time using the Open GL board that ships with most systems.

This render speed will increase over time, but if you find it too slow you can go into Marquee and turn off some of the quality settings under Render/Options. Avoid making large, soft, drop shadows if render speed is a problem on animations. You can set up a Marquee animation to render and then go back to editing, but you may find the rest of your system has slowed down too much to do much serious work.

There is much depth to Marquee and many people only scratch the surface. If you have the time you should explore the scripts and perhaps write some of your own. The Marquee scripts can be written in Python programming language, so if you have a repetitive task you could write a custom function to handle it (or pay your favorite Python programmer to do it for you). You can also import images that are larger than standard frame size and zoom and pan on the image. This is great for simulating motion-controlled camera moves. Since Marquee has keyframe and Bezier curves, you can get quite sophisticated motion on the images. And finally, Marquee can be used as a sophisticated multilayer 3D DVE if you are willing to spend the time to learn it.

Conclusion

The ability to layer, paint, and use plug-ins has given the Avid editor a whole range of tools and looks to create effects that look like they were made on much more expensive systems. Graphic looks continue to get more sophisticated and subtle, so taking the time to explore the Avid effects and the interoperability with third-party plug-ins and graphics programs will definitively pay off. Faster rendering and more real-time streams make more creative work possible in the same timeframe. Faster CPUs promise that more work can be done without dedicated hardware. Networks will allow users to distribute work and share media and, like the DS Nitris systems, distribute rendering to unused or dedicated rendering systems. Editors and designers will always continue to experiment and push the technology to the limits, and with the tools now becoming available for nonlinear editing with Avid, they have more choices than ever.

9

CONFORMING AND FINISHING

"The unfinished is nothing."

— Henri Frédérik Amiel

If editing is a finely balanced mixture of art and craft, then it could be argued that conforming is almost entirely craft, for conforming is all about precision. In this chapter we're going to delve into a very specific workflow for online conforms. There are variations and branches along the way, especially when conforming in high definition (HD), but fortunately the main thread works for nearly all conforms of an Avid offline.

Choosing the Finishing Resolution: Standard Definition

Before beginning your conform, you should select an appropriate resolution for media that you will be capturing in Avid online. Though sometimes the resolution is predetermined by the production or workflow requirements, the following guidelines will help you determine the best resolution to use for your online conform.

Though you can certainly do an online conform at a compressed resolution, I strongly recommend that you use uncompressed media for all of your standard-definition (SD) finishing. Finishing at a compression ratio of 2:1 is also acceptable under some circumstances and is sometimes preferred due to the reduction in disk space required and the fact that most SD conforms are delivered on either Digital Betacam, which itself uses a compression ratio of approximately 2.3:1, or IMX 50, which uses a compression ratio of approximately 3.3:1. I still prefer uncompressed files simply because using uncompressed material eliminates decompression time by the computer processing unit (CPU) and allows it to do more real-time effects processing. Rendering is time "wasted" in online—I prefer to do as little of it as I can get away with.

Avid Media Composer provides three different uncompressed media formats:

- *1:1 OMF:* 8-bit 4:2:2 $Y'C_BC_R$ uncompressed OMF media. This type of media is stored in OMFI MediaFiles folders on your media drives.
- *1:1 MXF:* 8-bit 4:2:2 $Y'C_BC_R$ uncompressed MXF media. This type of media is stored in Avid MediaFiles folders on your media drives.
- *1:1 10b MXF:* 10-bit 4:2:2 $Y'C_BC_R$ uncompressed MXF media. This type of media is stored in Avid MediaFiles folders on your media drives. As mentioned earlier in this handbook, only MXF can store 10-bit media on Avid.

Choosing the Finishing Resolution: High Definition

Avid systems support both uncompressed and compressed HD media. Depending on your finishing requirements, you may find that compressed media files are more than sufficient. However, if your project contains extensive keying and compositing, you may prefer to work with uncompressed HD media. All HD media are stored in the MXF format in the Avid MediaFiles folders on your media drives. The following HD media types are available:

- *1:1 10b HD:* 10-bit 4:2:2 $Y'C_BC_R$ full-raster (1920 × 1080 or 1280 × 720) uncompressed media.
- *1:1 HD:* 8-bit 4:2:2 $Y'C_BC_R$ full-raster (1920 × 1080 or 1280 × 720) uncompressed media.
- *Avid DNxHD:* Mastering-quality compressed HD media. The DNxHD family of resolutions provides both 8- and 10-bit, extremely high-quality, full-raster compressed media. Multiple compression levels are provided for each HD format. All of the DNxHD resolutions are considered mastering quality and an equivalent or higher data rate than either HDCAM, D5 HD, or DVCPRO HD. DNxHD compression and decompression are performed in real time in hardware on the Avid Symphony Nitris.

DNxHD media are named by their data rate in megabits/second instead of the compression level. As the data rate varies based on the HD format and frame rate, the specific numbering of DNxHD media varies from one format to another. For reference, the following resolutions are available in the 1080i/59.94 format:

- *DNxHD 220x:* 10-bit 4:2:2 $Y'C_BC_R$ full-raster (1920 × 1080 or 1280 × 720) 220 Mbsec compressed media. The compression ratio is approximately 5.7:1 for 1080i and 2.5:1 for 720p.

- *DNxHD 220:* 8-bit 4:2:2 $Y'C_BC_R$ full-raster (1920 × 1080 or 1280 × 720) 220 Mb/sec compressed media. The compression ratio is approximately 4.5:1 for 1080i and 2.0:1 for 720p.
- *DNxHD 145:* 8-bit 4:2:2 $Y'C_BC_R$ full-raster (1920 × 1080 or 1280 × 720) 145 Mb/sec compressed media. The compression ratio is approximately 6.8:1 for 1080i and 3.1:1 for 720p.

Audio Format Options

When conforming an online, 99 percent of the time you should work with 48-kHz audio samples using 24 bits per sample. The other 1 percent of the time you typically work at 48-kHz audio using 16 bits per sample. Why? Simply because digital decks all run at 48 kHz natively using either 16, 20, or 24 bits per sample. Working at 48 kHz means you can output baseband to an HD or SD deck via SDI (serial digital interface). Mastering at 44 kHz just isn't done anymore.

With that choice made, the only other audio mastering decision is whether to use OMF or MXF media to store your audio. Both provide identical audio quality so the decision usually rests on the preferred format for the audio engineer or department. Many audio postproduction departments still prefer to use OMF media, primarily because their equipment also supports it natively. If you are finishing your own audio, you can use either one. The primary difference for you is whether you'll store it in an OMFI MediaFiles folder or an Avid MediaFiles folder.

Delivery Requirements to Online

As an online editor you have very specific delivery requirements for the final program master. I feel strongly that if you have delivery requirements for your product, then you should place delivery requirements on those delivering the offline to you. Let's be honest. Online bay time is expensive. If all the elements are not delivered, you may not be able to complete the online and will instead have to wait, burning up the client's budget waiting for couriers to deliver missing elements or spend unnecessary time fixing elements that weren't delivered properly.

Offline Element Deliverables

Table 9.1 lists the basic delivery requirements expected from an offline editor. Some elements may vary, depending on project workflow.

Table 9.1 Offline Element Deliverables

Required Element	Notes
Offline project	Though only the final offline sequence is really needed for offline, it is helpful to have the entire project, especially if troubleshooting is required.
Digital cut of final offline	A digital cut of the offline is essential. If there are any questions about title placement, element alignment, or effect design, they can often be answered by examining the offline digital cut. *Note:* The digital cut should be laid off using sequence timecode to a timecoded tape format such as Beta SP or DVCPRO HD. Do not accept a VHS tape.
The final audio mix: • Audio media from offline • ProTools mix • Digital cut	The audio mix can be delivered in a variety of formats. The method of delivery will vary from project to project. Ensure that both you and the offline editor agree on the delivery method.
All required source tapes	The offline editor should double check that all tapes were packaged and sent to the online. A list of all source tapes should be included in each box of tapes, indicating which tapes from the production are included in each box. We'll discuss how to pull this list later in this chapter.
Nonstandard fonts used in offline	Any nonstandard fonts should be delivered to the online. Note that I'm not saying just a list of the fonts; I strongly recommend that the offline editor *include* the actual font files. Do not assume that the online facility owns the same fonts you do.
All online import elements	All graphics, animations, and audio used in the project should be delivered to the online. Additionally, these elements should meet a graphic delivery requirements spec. This spec is discussed in the next section.
List of AVX plug-ins used	If plug-ins were used in the offline, the online editor must know which ones were used so they can be made available in the online. We'll discuss how to pull this list later in this chapter.

Delivery Requirements for Import Elements

To ensure that the online goes smoothly, all graphics and animations should meet an online delivery specification. This spec should be given to the offline editor and all graphic artists, animators, and compositors who are providing elements for the project. Tables 9.2 and 9.3 list typical delivery requirements for graphics and animations.

Pulling a Source List

There are a number of techniques available in the Avid system to pull a source list, including using EDL Manager, as discussed in the Appendix. But the simplest method by far is using the `dumpsourcesummary` Console command. This method allows you to quickly and easily generate a list of all of your tape- and

Table 9.2 Standard Definition: Still Graphics

Aspect	Requirement	Notes
Frame size: 4 × 3 square pixel	648 × 486 (NTSC) 768 × 576 (PAL)	These are the preferred square pixel sizes for NTSC and PAL. 720 × 540 can also be used in some situations for both NTSC and PAL.
Frame size: 16 × 9 square pixel	864 × 486 (NTSC) 1050 × 576 (PAL)	These are the preferred sizes for NTSC and PAL. Note that the PAL size is wider than you might expect. This is due to the wide horizontal blanking region in 601 PAL frames.
Frame size: nonsquare pixel	720 × 486 (NTSC) 720 × 576 (PAL)	These are the native frame sizes for SD graphics.
Alpha channel	White on black	This is the standard used by all graphics, animation, and compositing packages. The alpha channel must be inverted on import.
Color mode	RGB	Other formats, including CMYK, indexed and grayscale, can cause import errors.
File format	TIFF (.tif), PICT (.pct), or PNG (.png)	These are the three most commonly used graphic formats. The PNG format allows for easy export of layered graphics out of Photoshop.
Additional Requirements for SD Animation and Video		
Field ordering	Even, lower field first (NTSC) Odd, upper field first (PAL) Even, lower field first (PAL DV)	Proper field ordering is critical.
Video level	RGB mapping	The other option, 601 mapping, should only be used when the source requires it (e.g., test patterns).
File format	Avid QuickTime codec	This is the preferred method of delivery.
Frame size (4 × 3 or 16 × 9)	720 × 486 (NTSC) 720 × 576 (PAL)	Avid QuickTime codec requires the full ITU-R BT.601 frame size.
Resolution	Uncompressed (1:1)	It is strongly recommended that all SD graphics are uncompressed.

RGB = red, green, blue; CMYK = cyan, magenta, yellow, black.

file-based sources in one pass. You can either copy and paste the resultant data out of the console into a text file or have the console send the data directly to a text file.

To generate a source summary:

1. Load the sequence to be onlined into the Record monitor.
2. Press Ctrl/Command+6 to open the Console.

Table 9.3 High Definition: Still Graphics

Aspect	Requirement	Notes
Frame size: 1080line	1920 × 1080	HD formats natively use square pixels.
Frame size: 720line	1280 × 720	
Alpha channel	White on black	This is the standard used by all graphics, animation, and compositing packages. The alpha channel must be inverted on import.
Color mode	RGB	Other formats, including CMYK, indexed, and grayscale can cause import errors.
File format	TIFF (.tif), PICT (.pct), or PNG (.png)	These are the three most commonly used graphic formats. The PNG format allows for easy export of layered graphics out of Photoshop.
Additional Requirements for HD Animation and Video		
Field ordering: 1080 interlaced	Odd, upper field first	Interlaced HD uses field ordering that is opposite of NTSC. If using a progressive HD resolution (1080p or 720p), field rendering should not be used.
Video level	RGB mapping	The other option, 709 mapping, should only be used when the source requires it (such as luma key elements, animated test patterns, preserved highlights, and so on).
File format	Avid DNxHD QuickTime or animation codec	Avid DNxHD is preferred for RGB animations. If an alpha channel is required, either DNxHD or Animation is acceptable.
Resolution	To match project requirement	Though uncompressed HD is preferred, DNxHD is suitable for most applications and imports much more quickly.

RGB = red, green, blue; CMYK = cyan, magenta, yellow, black.

3. Choose "Open Log File" from the Console's Fast menu.
4. Type a name for the source summary dump and press Enter.
5. Type `dumpsourcesummary` into the Console and press Enter. The source list will be displayed to the Console.
6. Choose "Close Log File" from the Console's Fast menu.

The source summary provides you not only with the list of every tape-based source, but also the project it was originally logged in on. This is a critical piece of data if you have duplicate source names. If you see any duplicate source names, these tapes should be flagged as you may be forced to eye-match a shot or two from one of these tapes to differentiate the two, especially if they share common timecode.

For the file-based sources, notice that the source summary includes the complete path for the imported files. This can be a real time-saver when it comes to properly gathering up these files for the online.

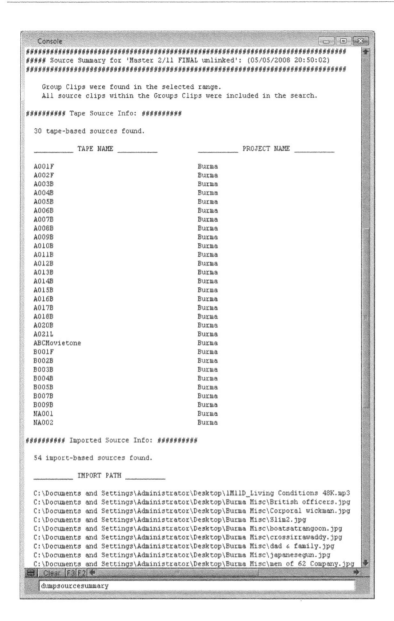

Unfortunately, a list of the fonts used in either the Title Tool or Marquee cannot be quickly gathered via the Console at this time. You or the offline editor will need to keep careful notes of the fonts used.

Pulling an Effect Plug-In List

After you pull a source list you should also pull a plug-in list. *It is critical that this list be generated on the offline system with all used effects installed.* This is due to the fact that not all plug-in effects store their name in a human readable form. If the plug-in is installed when this list is generated, the command will quickly parse the installed plug-ins and extract a human readable name from the plug-in itself. Otherwise all the command can return is the effect's hash code.

To generate a source summary:

1. Load the sequence to be onlined into the Record monitor.
2. Press Ctrl/Command+6 to open the Console.
3. Choose "Open Log File" from the Console's Fast menu.
4. Type a name for the source summary dump and press Enter.
5. Type dumpfxsummary into the Console and press Enter. The source list will be displayed to the Console.
6. Choose "Close Log File" from the Console's Fast menu.

> Since the original publication of this book a complete user interface has been provided for these console commands within the "Clip > Get Info" command. You can now also directly output these reports to a text file.

The Online Project

Though a sequence could certainly be recaptured in the original offline project, there are significant advantages to using a clean project for the online. By creating an online project you can easily configure the Project settings and eliminate any possible errors or problems created by Project settings created and used in the offline.

I recommend creating a new online project for each job being conformed. This ensures that the project format is properly configured and any unique configurations for one specific project do not carry over to the next.

Online Project Settings

There are a number of specific setting configurations that I recommend when building an online project. Some of these settings will vary depending on the nature of the online, but generally using these settings will help the online go as efficiently as possible.

Audio Project Settings (Main Tab)

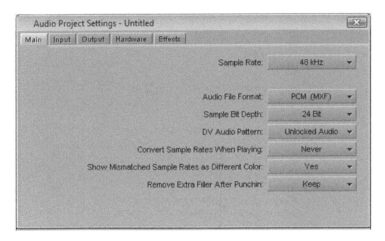

- *Sample rate and sample bit depth:* Set as required by the project. In most cases, if you are mastering to Digital Betacam, HDCAM, or a similar format, these should be set to 48 kHz and 24 bits, respectively.
- *Audio file format:* Both OMF and MXF media formats are available and can be freely mixed in a sequence. Let's look at the three available options.
  - *WAVE (OMF):* The media are packaged using the Wave format, which is readable by nearly all Windows applications that support sound. These media are stored in the OMFI MediaFiles folders on your media drives.
  - *AIFF-C (OMF):* The media are packaged using the Audio Interchange File Format, which is readable by most computer sound applications. These media are stored in the OMFI MediaFiles folders on your media drives.
  - *PCM (MXF):* The media are packaged using the industry-standard Pulse Code Modulation format. These media

are stored in the Avid MediaFiles folders on your media drives.

- *Convert sample rates when playing:* Set to "Never" for all online work. Converting sample rates on-the-fly may produce a lower-quality result.

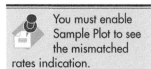

You must enable Sample Plot to see the mismatched rates indication.

- *Show mismatched sample rates as different color:* Set to "Yes." This enables you to visually identify audio clips with the incorrect sample rate in the timeline.

Capture Settings (General Tab)

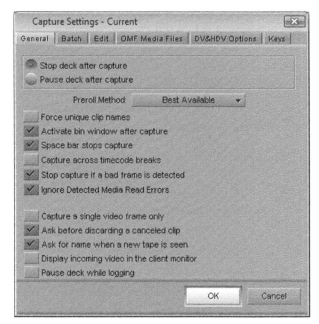

Preroll Method: This menu controls how the Avid editing system controls the deck during preroll. Four options are available:

- *Standard Timecode:* Instructs the editing system to always seek the preroll point by direct access. This is done by subtracting the specified preroll duration from the In point timecode. If the preroll timecode does not exist, the clip is not captured and an error is reported.
- *Standard Control Track:* Instructs the editing system to always seek the preroll point by control track access. This is done by first seeking the In point then switching the deck to Control Track Offset mode and rolling backwards by the specified preroll duration. If sufficient continuous control track does not exist prior to the In point, the clip is not captured and an error is reported.
- *Best-Available Control Track:* Instructs the editing system to seek first using Standard Control Track. If a control track

break is encountered, the system will shorten the preroll to the amount of continuous control track available. If there is not enough control track prior to the In point, the clip is not captured and an error is reported.

- *Best Available:* This option instructs the editing system to try the following options, in order:
  - Standard Timecode
  - Standard Control Track
  - Best-Available Control Track

The default option, Best Available, provides the greatest chance to capture the material but can take longer, in some instances, than other methods if chosen directly. For these reasons we recommend doing the following (in conjunction with other Capture settings options):

- Choose *Standard Timecode* first when capturing material previously captured in the offline. This is the fastest capture method.
- After batch capturing a reel, if clips still remain offline, change the Preroll Method to *Standard Control Track* to capture the remaining clips.

Capture Settings (Batch Tab)

- *Optimize for batch speed:* Enable this option to potentially speed up the capture process. If this option is active, and the distance between the Out point of one clip and the In point of another is five seconds or less, the deck will not pause between the two clips and will roll forward to the next clip. Since this is not an uncommon occurrence when capturing from a decomposed offline sequence, enabling this option can save a significant amount of time (and wear on the deck's transport) that would otherwise be spent seeking and prerolling.
- *Switch to emptiest drive if current drive is full:* Though this option may be useful in some local storage–only configurations, it is generally better to manually manage your storage and specify to the system which disks or partitions you wish to use when capturing.
- *Eject tape when finished:* If enabled, the tape will be ejected after the last clip from that reel is captured. Ejecting the tape is a useful prompt for the editor, assistant editor, or tape operator that the system is ready for the next tape. This option is particularly helpful when working with machine rooms.
- *Log errors to the console and continue capturing:* It is strongly recommended that this option is enabled when batch capturing. If an error, most likely a coincidence error, is reported by the deck, the system will note the error

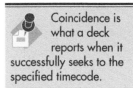

Coincidence is what a deck reports when it successfully seeks to the specified timecode.

and proceed to the next clip. If this option is not selected, the system will pause the capture and display a dialog box every time an error is encountered.

Deck Settings

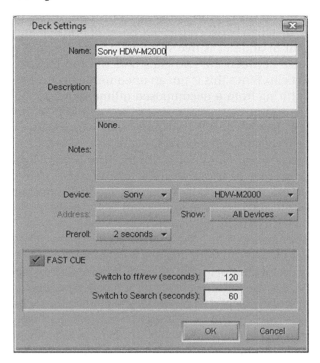

Accessing Deck Settings

The Deck settings dialog box is accessed via the Deck Configuration setting as follows:

1. Open the Deck Configuration setting. The current deck patch will be displayed, showing the control port and deck. If this dialog box is empty, you should first connect the deck and select "Auto-configure."
2. Double-click the deck on the right side of the patch to open the Deck settings.

- *Preroll:* Sets the deck's default preroll duration. This value defaults to five seconds for most decks. Well-maintained modern digital decks, such as the Sony DVW-A500, Sony HDV-F500, Sony HDW-M2000, and Panasonic AJ-HD3700, can get to speed and lock in less than one second. Reducing this preroll duration to one or two seconds can slice minutes or even an hour or more off the time required to recapture a long-form program with a lot of edits.
- *Fast cue:* Instructs the system when to switch to FF/REW on a deck (to speed up access to the requested timecode). This option should be on for all standard baseband decks. We've found, however, that leaving this option on for baseband capture from some XDCAM decks will dramatically slow down access to the material. Disable it for those XDCAM decks.

Media Creation (Drive Filtering and Indexing Tab)

- *Filter Out System Drive/Filter Out Launch Drive:* When disabled, these options allow the system to digitize to the drive containing either the operating system or the Avid application. Newer Avid systems ship with very large internal drives, which some offline editors use for audio or lower-resolution media storage. These options should be enabled on finishing systems.

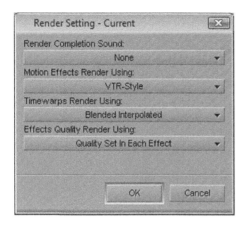

Render Settings

- *Motion Effects Render Using/Timewarps Render Using:* These two options allow you to specify a default rendering method for motion effects and timewarps. Offline editors often choose the most expedient rendering method (e.g., Duplicated Field) instead of the appropriate method for the online. These two pop-up menus allow you to specify the method that will be used in the final sequence when it is rendered. These two options are usually set to VTR-Style and Blended Inter-polated, respectively.

Necessary Equipment for Online Suites

Your Avid suite is now an online suite if you plan to finish there. By purchasing the highest resolutions and the fastest drives, you have not completed this transformation! Online suites have some important components that must be present. There must be a high-quality third monitor to view reference video. This must be an engineering-quality monitor. You need one that can be adjusted through standard adjustment procedures like using the blue bars setting. The monitor must hold that calibration in a stable fashion over time. This generally means more money for the monitor, but if you ever get into a dispute about color or brightness, this monitor is your absolute. It is the end of an argument and the last word on "what it really looks like." Don't put your faith in cheap equipment—this finished image is your reputation!

Some people like to have a consumer-type monitor in the room as a low-end reference to "see what it will look like at home." Be careful of this since it is very hard in NTSC to get certain colors to look exactly the same on both monitors. When your client insists that the yellow on the low-end monitor must match your high-end monitor, you have a frustrating no-win situation. This is why many online tape suites have only one color monitor and everything else is black and white.

There are very precise color-measuring devices that allow you to adjust monitors to match more closely. After such an adjustment, a video monitor should be left on all the time to minimize the drifting that occurs with warming up and cooling down. If you need to adjust a color monitor, always wait until it has been on for several minutes, longer for older monitors.

It is also nearly impossible to get the computer monitors to match up to the video monitors. They are two very different types of monitors and, even though you can adjust the computer monitors, you cannot adjust hue through hardware adjustments. Color-matching extensions give you more control than the front panel buttons. Many professional graphics programs include these gamma-adjusting extensions, but you will never get the monitors to match completely. You can also use an automated device that can run test patterns and self-adjust. Using one of these calibration devices (e.g., an X-Rite Hubble) can help you match the computer monitors with your video monitors. Do not let clients pick a final color on the computer monitor. They must have final approval on the well-adjusted, carefully lit, high-end video monitor. And yes, it will look different at home.

You will need external waveform and vectorscopes as independent references for all signals. Use them to monitor input, output, and dubs. If you can set up a patch bay or router, everything should go through these scopes. You cannot consider yourself an online room until you have waveform and vectorscopes.

More and more people are deciding that having a color corrector in the suite gives them the extra protection and ability to deal with sources that have difficult color problems. It also gives you an extra tool to make sure that sources shot in very different locations really match up when they are next to each other. Since there is no hue adjustment for PAL and component NTSC, and no adjustment at all for serial digital input, the color corrector gives you the level of control you occasionally need. It can also give you some illegal levels if you are not careful! If a black looks really rich and the shadows have that deep dramatic look you really want, check that the black level never goes too low.

You need studio-quality speakers, and they must be mounted far enough apart so that you can listen critically for stereo separation. If you can afford it, get a good compressor/limiter. This takes your audio levels and gently compresses the loudest parts of the sound so that they don't distort. This means you can have your overall sound louder and not worry about the occasional spike in sound level. You can use it during both the capturing and the output or you can use an AudioSuite™ plug-in before you output. However, if you are sending your audio to a sound designer to finish, don't compress it. Leave the sound alone except to adjust levels and make sure nothing is distorted.

Onlining and Offlining on the Same Machine

In some situations, the online may be done on the same system as the offline. If this is the case, you will likely need to make room on your media drives before you can begin recapturing at the online resolution. There are two common methods for creating the needed storage space for your online conform. One method is fairly easy to perform and involves a general purging of all of the project's extraneous media from the system. A second method is slightly more complex, but allows you to keep an "offline-quality" version of the sequence online throughout the conforming process.

Method 1: Delete the Offline Video and All Unused Media

Use this method if you want to keep all the audio media from the offline and only delete the offline video media. This method will also allow you to keep any final online-quality media that are used in the sequence, such as uncompressed imported graphics. This procedure should only be done before you begin recapturing the media at the final online resolution.

Part 1: Delete the Offline Video Media

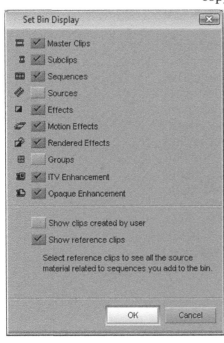

1. Create a new bin, name it "Purge Offline Media," and copy the final offline sequence into it. If there is more than one version to be onlined, copy all of the versions into this bin.
2. Activate the new bin and select "Bin > Set Bin Display."
3. Select the following:
   • In the upper section: "Rendered Effects."
   • In the lower section: "Show Reference Clips."
4. Deselect the following:
   • In the upper section: "Sources, Groups."
   • In the lower section: "Show clips created by user."
   • The other items can be left in their original state.
5. Click OK. The bin's view changes to display only the clips and rendered effects (precomputes) referenced by the sequence.
6. Press Ctrl+A to select everything in the bin and press Delete on the keyboard. The Delete Media dialog box appears.
7. In the upper section, select both of the "Delete associated media file(s)" options.

8. In the lower section, deselect "Audio" and any "online" resolutions such as "Uncompressed."
9. Click OK to delete the media.

Be sure you have the right options selected and deselected as there is no delete confirmation dialog box after you press OK.

All of the original offline video media files are deleted, including rendered video effects. The next step is to delete all the media not required by the offline sequence(s).

Part 2: Delete Unrelated Video and Audio Media Files

To finish creating your storage space you now need to delete the project's unused video and audio media files. This is easily done via the Media Tool.

1. Select "Tools > Media Tool" to display the Media Tool Display dialog box.
2. Select the following options:
   • The media drive(s) to which the offline media were digitized.
   • Current project.

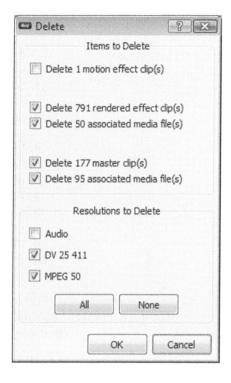

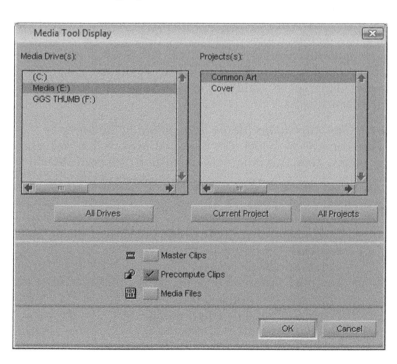

- You may also need to select other projects if any clips originated in a different project.
- Master clips and precompute clips.

3. Click OK.
4. Select the Purge Offline Media bin and press Ctrl+A to select all the remaining reference clips.
5. Select "Bin > Select Media Relatives." The media the sequences require are selected in the Media Tool.
6. From the Media Tool Fast menu, choose "Reverse Selection." All of the items selected in the Media Tool are unrelated to the offline sequence and can be deleted.
7. Press the Delete key.
8. Select checkboxes for any video, audio, or precompute tracks available.
9. Click OK.
10. If desired, select "Bin > Set Bin Display" and reset the view to display the sequence(s) and hide the reference clips.

The only media files that now remain in your project are the required audio media files and any online resolution material that was edited into the offline sequence.

Method 2: Keeping the Offline Edit Available

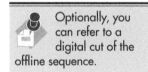
Optionally, you can refer to a digital cut of the offline sequence.

If the offline was done on the same system, you may prefer to have an offline-quality version of the sequence available to refer back to as a reference while you conform the final version of the show. This method allows you to do just that.

The following two procedures will help you maximize the amount of storage available. Use Part 1 to generate a significant amount of free space. Add Part 2 to help create even more storage space if you need it.

Part 1: Clearing Excess Media through Media Relatives

First, remove all media files unrelated to your final sequence from your system. This procedure is similar to the one in Method 1.

1. Select "Tools > Media Tool" to display the Media Tool Display dialog box.
2. Select the following options:
   - The media drive(s) to which the offline media were digitized
   - Current project.
   - You may also need to select other projects if any clips originated in a different project.
   - Master clips and precompute clips.
3. Click OK. The Media Tool appears.
4. Select the final offline version(s) of the sequence in your bin.

5. Choose "Bin > Select Media Relatives." All of the media relatives are highlighted in the Media Tool.
6. From the Media Tool Fast menu, choose "Reverse Selection." All of the items selected in the Media Tool are unrelated to your final sequence and can be deleted.
7. Press the Delete key.
8. Select checkboxes for any video, audio, or precompute tracks available.
9. Click OK.

The only media files associated with your project now are the video, audio, and precompute effect media files required to play back your sequence.

If you discover, after eliminating the excess media files, that you require even more storage space, you may want to continue on to Part 2.

Part 2: Using Consolidate to Create Additional Storage Space

This stage "trims the fat" off of the original media files kept for the offline. You use the Consolidate command to create smaller copies of the media files used to play back the sequence.

When you consolidate a sequence, the system finds the media files (or sections of media files) required for playing the sequence. The system copies these media files (or parts of media files) to the target disk that you specify. The process is optimized to create as few new media files with as little digitized video and audio data as possible. Therefore:

• The system creates new media files and master clips that are shorter sections of the original media files and master clips. The new media files and clips are identified by the ".new" extension in their names.
• The system breaks the links that connected the sequences to the original media files. The new consolidated media files are not linked to any subclips or sequences except the sequence that you consolidated.

To create additional storage space by consolidating your sequence:

1. Select the final offline sequence.
2. Choose "Clip > Consolidate/Transcode."
3. Select "Delete original media files when done."
4. Deselect "Skip media files already on target drive."
5. Select a target disk from the Video Target Drives list. The system will alert you if the disk does not have enough storage space.

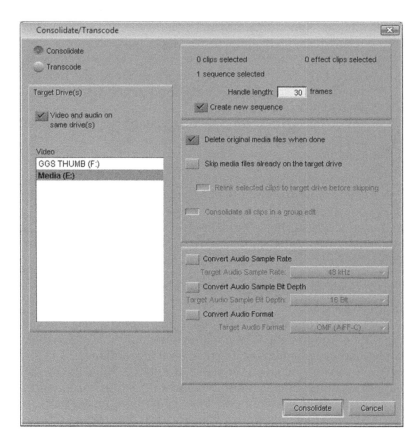

6. Choose a handle length. The amount of handle added is in addition to existing media required for transition effects. A handle length of ten frames is usually sufficient to allow some flexibility in the online conform while saving valuable storage space for the online-quality media.

7. Click Consolidate to begin the consolidation process.

Preparing to Recapture

Before you can begin recapturing, you need to prepare the sequence. Recapturing a sequence on Avid Media Composer or Avid Symphony should be as fast and as efficient as possible. But, you also need to have significant control over the process in case you need to make changes later.

Using the Online Project

As mentioned previously there are significant advantages to using a clean project in the online instead of the project used in the offline.

1. Copy the bin containing the final offline sequence into the online project.
2. Launch Avid Symphony Nitris and open the online project.
3. If necessary, rename the offline bin, indicating that it contains the offline version (e.g., "CommonArt—Locked Offline Edit").
4. Create three additional bins and name them as follows:
   - Online MOS (e.g., "CommonArt—Online MOS").
   - Decomposed clips (e.g., "CommonArt—Decomposed Clips").
   - Online final (e.g., "CommonArt—Online Final").

These bins will be used to store versions of the sequence as it goes through the online process.

Decomposing the Sequence

The Decompose function is the first half of the batch digitizing process. Decompose breaks your sequence into individual offline master clips. These master clips represent only the portion of the clip that was used in the sequence, so recapturing is very efficient. Because individual clips are created, you can easily sift and sort by tape prior to digitizing and identify and isolate clips that caused digitizing errors during the batch digitize.

Though it is possible to recapture a sequence without decomposing it, doing so has several disadvantages. For example, you do not have control over the order of tapes during the recapture process nor are you able to flexibly abort and resume the batch digitize process.

When you decompose a sequence:

- The system creates new master clips that are shorter sections of the original master clips and identifies them with a ".new" extension in their names.
- The system breaks the links that connected the sequence to the original master clips and creates new links to the ".new" clips.
- The ".new" clips created through the decompose process have never been digitized and have no link to any media files. The only link they have is to the sequence from which they were created.

The most common onlining scenario is to recapture only the video. The audio will either be brought forward from the offline, or provided as a separate element for integration in the online. The most efficient approach in this scenario is to decompose and then digitize a video-only version of the sequence.

Part 1: Create a Video-Only Version of the Sequence

1. Open the offline and MOS bins.
2. Duplicate the offline sequence and move the duplicate to the Online MOS bin.
3. Close the bin containing the original offline sequence.
4. Rename the duplicated sequence to indicate that it is a video-only version; for example, you might want to append "ONLINE MOS" to the end of the name.
5. Load the sequence into the Record monitor.
6. Select all audio tracks.
7. Deselect all the video tracks.
8. Press Delete to remove the audio tracks from the sequence.

Part 2: Decompose the MOS Version

Decompose is also available via the contextual pop-up menu.

1. Select the duplicated sequence in the bin and choose "Clip > Decompose."

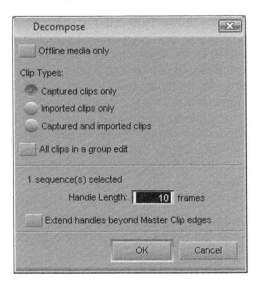

2. Deselect "Offline media only." If none of the offline media is available, you can leave this option selected.
3. Select "Captured clips only" to decompose only the captured video clips.

If you have also used XDCAM media you should select "Captured and imported clips" instead.

4. If available, deselect "All clips in a group edit." This option is only available if you are working with grouped or multi-grouped source clips. Groups are typically used when working with multicamera material. If selected, this option will decompose not only the used camera for each edit, but the matching footage for all additional cameras. Needless to say, this will create substantial unnecessary

material and dramatically increase the time required to recapture the sequence.

5. Enter a suitable handle length for each decomposed clip. Usually a handle of five to ten frames is sufficient when onlining. Since the handle is added to the head and tail of every edit in the sequence, a larger setting for handle can significantly increase the amount of media to be digitized and slow down the online conform.

6. Deselect "Extend handles beyond Master Clip edges." When this option is disabled, the system won't extend handles beyond the start and end of the master clip, preventing the system from creating a master clip that may contain nonexistent timecode. This is especially important for field tapes shot with the camera's timecode generator set to "Free Run" (for generating time-of-day timecode).

7. Click OK. A dialog box appears.

8. Click OK.

As the sequence is decomposed, new master clips will be created in the Online MOS bin for each video edit in the sequence. All links to the original offline master clips are broken and the decomposed sequence is taken offline.

Part 3 (Optional): Modifying the Decomposed Clips

There is one thing to consider before moving on to the next phase. After you have recaptured at the online resolution, will any changes be required? If so, it may be difficult since the video and audio come from different clips.

For example, if you match-frame a video edit from your final sequence, it will match to a video-only (MOS) clip. This can be somewhat disconcerting to the client in the online bay. To avoid this problem, you can add audio tracks to the decomposed MOS clips. This will not increase digitize time and will not significantly increase storage requirements.

If you edit this audio into the online sequence, you will need to check and adjust the mix to integrate the new audio into the mix.

1. Select all the ".new" clips that resulted from the decompose process.

2. Choose "Clip > Modify."

3. Choose "Set Tracks" from the pulldown menu.

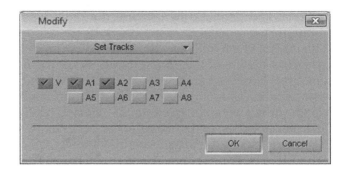

4. Enable the appropriate audio tracks (usually A1 and A2) and click OK.

When these clips are recaptured at their higher online-quality resolution, they will be captured with video as well as audio media.

Be sure the audio sample rate you have set in the Audio Project setting matches the audio sample rate used in the audio mix. If the sample rates do not match, the audio for the ".new" clips will play back as silence when edited into the final sequence. (Remember previously we told the system *not* to play mixed sample rates in real time.)

Pulling a Source List

The offline editor should have generated a source list and confirmed that all tapes were sent to the online. Occasionally, though, a tape is overlooked. You should also pull a source list using the dumpsourcesummary Console command and confirm that you have all required tapes before you start capturing.

Flagging Clips from Duplicate Tape Names (Optional)

If the offline editor borrowed master clips from older projects it is possible that the online sequence will include duplicate tape names. These duplicates can cause real problems in the online.

For example, if there were two different tapes named "001," when the Avid system asks you to load reel 001, would you know which 001 it wanted? If you loaded the wrong 001 and the clip's timecode existed on the tape, the deck would digitize whatever footage it found at that timecode, even if it wasn't the right material!

Though many onlines will not have this problem, it is a good idea to check for duplicate identifications (IDs) and then flag the suspect clips.

1. Load the decomposed sequence into the Record monitor.
2. If you haven't already, use dumpsourcesummary to dump a list of the sources. This list is sorted by tape name then by

project. Any duplicate tape names should appear next to one another in the list with their associated project's listed to the right.

3. If you see any duplicate tape IDs, print the summary then highlight them in the printout of the summary. I strongly recommend flagging these clips in the bin so the tape they're from is called out. Two great methods for this are using a clip color or adding asterisks to the head of the clip name. First let's set up the bin so it matches the summary printout.

4. Highlight the bin containing the decomposed sequence.

5. Select "Bin > Headings" to display the Bin Column Selector dialog box.

6. Press the All/None button twice to deselect all the columns.

7. Select both the Tape and Project columns then press OK.

8. Save this view as "Tape/Project."

9. Select the Tape column heading then hold the Shift key down and also select the Project column heading.

10. Press Ctrl/Command+E to sort the view by tape and then by project.

11. Scan through the bin, looking for duplicate tape names.

12. Flag the duplicated ID clips so they can be identified during the online.

Solving Duplicate Tape Names

Unfortunately, there is no easy way to fix duplicate tape name problems in the online. But, duplicate IDs can be avoided in the offline by using proper tape naming conventions. Many offline editors have gotten into the bad habit of naming the first tape used in a project "001," the second tape "002," and so on. Though this habit worked fine in the linear world where each sequence (edit decision list) was an island, this approach does not work in the nonlinear world where it is easy to borrow material from other projects and sequences.

A good strategy to adopt in the offline is to assign a unique tape name to every single tape across all projects. For example, the offline editor can simply append the show ID to the tail of the tape number. Using the previous illustration, this approach would result in two tape names, "003-618" and "003-704." Another approach is to use a tape library system and assign a unique number to each tape that enters a facility.

Duplicate tape names slow down the online process and introduce the risk of the wrong footage being used in the sequence. If you are working with an offline editor that uses duplicate tape names explain to him or her and the producer the problems, additional time, and additional costs this is creating in the online. Encourage them to adopt proper tape naming conventions for future programs.

Assigning Source Ownership Using Avid Log Exchange

To make media management and cleanup easier, you want all the clips in the project to belong to the online project. Currently, all of the clips belong, at best, to the offline project. In reality, they may belong to several projects, depending on how the logging was done or if clips were borrowed from other projects.

The Modify command can change the source ownership of a clip, but that method is tape name based and can be quite time consuming. Here is a fast method using an Avid Log Exchange (.ALE) file.

If you are conforming a film project, you will need additional columns for the film-related metadata. Choose the Film bin view and add the Tape column. Then save this view as ALE.

1. Open the Decomposed Clips bin.
2. Select all of the decomposed clips in the Online MOS bin and move them to the Decomposed Clips bin.
3. Display the following headings in the Decomposed Clips bin:
   - Start
   - End
   - Tracks
   - Tape
4. Save this view as ALE.
5. Select all clips in the bin.
6. Choose "File > Export."
7. Open the Export settings, and choose "Avid Log Exchange" from the pop-up menu.
8. Click Save As and name the Export setting "ALE Export."
9. Save the exported file to your drive and close the Decomposed Clips bin.
10. Select the Online MOS bin.
11. Select the ALE bin view.
12. Choose "File > Import."
13. Select the ALE file you just created and press OK to import the file and create new clips. Notice that the Start, End, Tracks, and Tape settings have not changed. If you add the Project heading to the bin, you will see that all of the clips are now assigned to the online project.

Linking the Sequence to the New Clips

The sequence is still linked to the original decomposed clips, not the newly created clips. Before we begin to capture from tape we need to link the sequence to these new clips.

1. Select the Online MOS bin and press Ctrl+A to select everything in the bin.
2. Select "Clip > Relink."
3. Choose "Relink all nonmaster clips to selected online items" and "Allow relinking to offline items."
4. Deselect "Create new sequences."

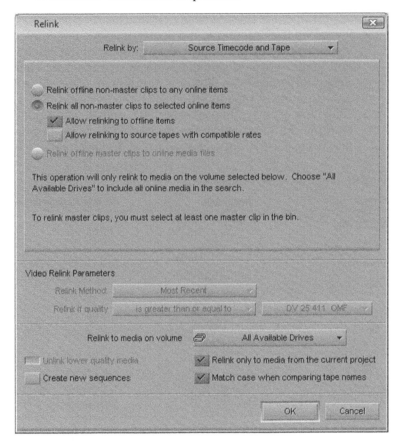

5. Click OK.

The sequence is now linked to the new clips. You can confirm this by using the Show Media Relatives command.

Calculating the Space Required for the Recapture

Before batch digitizing, you should check the space required for the online media.

1. Select all the clips in the Online MOS bin.
2. Ensure that the sequence is not also selected.
3. Press Ctrl+I. The Console opens, and the duration of the decomposed clips is listed.

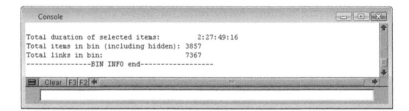

4. Round up the duration to the nearest minute and convert the total duration to minutes. For example, for the illustration here, we would round 2:27:49:16 to 2:28:00:00. Then we would convert the duration to minutes, resulting in a total duration of 148 minutes.
5. Table 9.4 lists typical SD and HD online resolutions. Multiply the duration by the desired value to determine the space, in GB or MB, required to capture the media.

Table 9.4 Resolution Storage Requirements

Format	Resolution	Storage Consumption
NTSC or PAL	1:1	1.22 GB/min
	2:1	0.53 GB/min
1080i/59.94	1:1	8.68 GB/min
	DNxHD 220	1.54 GB/min
	DNxHD 145	1.00 GB/min
1080i/50	1:1	5.79 GB/min
	DNxHD 185	1.28 GB/min
	DNxHD 120	0.85 GB/min
1080p/23.976	1:1	5.56 GB/min
	DNxHD 175	1.22 GB/min
	DNxHD	0.81 GB/min

For example, if we were recapturing this sequence in HD using DNxHD 220, we would multiply 148 by 1.54 and determine that we needed roughly 227 GB to recapture this sequence. As a point of comparison, if we were to recapture the sequence using uncompressed HD, we would need 1285 GB!

Batch Capturing

The most efficient method to acquire the newly created clips is to batch capture them tape by tape. You can ensure this process goes as efficiently as possible by properly configuring the system for capture. As we mentioned previously, there are several Capture and Deck configurations that should be changed from their defaults for the best performance during a batch capture. It is also wise to confirm that you are capturing at the correct resolution and to the appropriate media type before you begin the capture. Use the following lists as a "preflight" settings checklist.

Capture Settings:

- *Preroll method (General tab):* Choose Standard Timecode when capturing from most tapes. This is the fastest capture method but it will generate errors if coincidence cannot be reported by the deck.
- *Optimize for batch speed (Batch tab):* Minimizes deck pausing and recueing. If two clips to be captured are five seconds or less apart on the tape, the system will continue rolling between the two clips instead of stopping and prerolling to the next clip.
- *Log errors to the console and continue capturing (Batch tab):* If an error is encountered during the capture process, the system will log the error to the console and move on to the next clip.
- *Eject tape when finished:* Instructs the system to eject a tape after capturing the last clip. Though this option isn't as important as the others, it can still save valuable time and serve as a prompt to insert the next tape.

Deck settings:

- *Preroll:* Set the preroll to a more appropriate value. For most modern decks a preroll of one or two seconds is all that is required.

Media Creation Tool:

- *Video file format (Media-Type tab):* Confirm that you are set to the appropriate media format. If you plan to capture with 10-bit samples, you must be set to MXF or the 10-bit resolutions will not be available. Remember that OMF media will be stored in an OMFI MediaFiles folder while MXF media will be stored in an Avid MediaFiles folder.
- *Video resolution (Capture tab):* Confirm that you are set to the appropriate resolution. Even though this can be set in the Capture Tool, it is a good idea to set it in advance.

Audio Project settings:

- *Sample rate and bit depth (Main tab):* Confirm that you are set to the required rates for the program's delivery requirements. If you will be capturing audio via SDI you should ensure that the sample rate is set to 48 kHz.
- *Audio file format (Main tab):* Confirm that you are set to the appropriate media format. Remember that both Wave and

AAIF-C media will be stored in an OMFI MediaFiles folder while PCM media will be stored in an Avid MediaFiles folder.

- *Input source (Input tab):* If you are capturing audio from tape, be sure to set the appropriate audio input. Even though this can be set in the Capture Tool, it is a good idea to set it in advance.

Bin configuration: We recommend that you display, at a minimum, the following bin headings:

- Tape
- Start
- Duration
- Tracks
- Offline
- Video
- Drive

You should then sort by tape and then start timecode. This allows you to organize and, if another deck is available, preroll the tapes to the first section you will capture from.

You can now digitize the new master clips:

1. Choose "Tools > Capture" or "Bin > Go to Capture Mode" to open the Capture Tool.
2. Select all the clips from the first tape to capture.
3. Choose "Clip > Batch Capture" to begin the capture process.

The system will prompt you to enter the first tape. Once you enter the tape the prompt will automatically disappear.

As the clips are captured, their offline status will change in the bin, making it easy to see which clips have been captured and which remain to be captured. If there are some clips that remain uncaptured when the last clip on a tape has been reached, then the system encountered an error during the clip's capture. Though other possibilities exist, most of the time this is due to a coincidence error on the deck.

1. Select the Capture Tool and press Ctrl+= to open the Capture settings dialog box.
2. Select the General tab and change the Preroll Method to "Standard Control Track," then click OK.
3. Select the clips that were not captured and choose "Clip > Batch Capture" to begin the capture process.

If, after the last clip is reached, there are still clips that remain uncaptured, open the Console and check the error message for those clips. It is certainly possible that a system error prevented their capture, but this is less likely if some of the clips from the

 Avid Media Composer 3.0 contains a prebuilt *Capture* bin view. Remember that you must create a new user when you install version 3.0 in order to include the new view in your user setting. (Remember to *always* create a new user when upgrading to a new version!)

 Since prerolling by control track requires two seeks, one to the start timecode and the other to the preroll point, I strongly recommended that you only use this setting when recapturing clips that did not capture with a standard timecode preroll.

tape were captured. One possibility is that there are timecode breaks on the source tape and the timecode decrements at a break. Few decks can correctly handle this scenario and usually report an error. You may need to manually locate the proper section on the tape then try capturing the clips again.

Analog Source Format Considerations

If you are capturing from an analog source, you may want to use the analog deck's TBC (Time Base Correction) controls—or an outboard TBC if capturing from a consumer source such as a VHS tape—to calibrate and adjust the video signal on each tape.

Clip-by-clip calibration via the deck's TBC should only be done if the video signal exceeds the limits of the ITU-R BT.601 standard (e.g., if the luminance exceeds 108.4 IRE). Otherwise, it is usually easier to correct for signal variations using color correction.

To use the TBC to prevent signal clipping:

1. As soon as you notice the signal is being clipped, use the TBC to lower the signal so it is no longer clipped. If you are digitizing a short clip, it may not be possible to adjust the signal before the end of the clip is reached. In that case, abort the digitize, adjust the TBC, and resume the digitize.
2. Click the Trash icon in the Digitize Tool. This aborts the batch digitize process for the current clip. A dialog box appears asking if you wish to retry digitizing the currently selected clip, continue batch digitizing the next clip in the list, or abort the batch digitize entirely.
3. Retry the current clip. As the system digitizes the clip again, confirm that the signal is no longer being clipped. Typically, the only corrections you'll need to make are with gain and saturation controls. Black levels can also slip and require correction, but that is less common. If you suspect the black level has slipped, you can easily correct it by examining the area of blanking within the video signal. The blanking area should measure 0 IRE.
4. After you have successfully recaptured the clip, read just the TBC to its original setting and continue digitizing.

External TBC adjustments are not stored by the deck. Therefore, it is your responsibility to recalibrate all sources manually during the recapturing process every time. This can be a potential drawback if you have to recapture a show months later for a second output. The second calibrations may not be identical to the first and the two shows may look slightly different. However, some external TBC controls have scene memory and can be used to store and recall specific TBC settings.

Before capturing from the next tape open the Capture settings and change the Preroll Method back to "Standard Timecode."

Integrating the Audio Mix

When you have finished recapturing the video, you need to reintegrate the audio back into the sequence. The audio can come to the online in several different formats:

- A ProTools mix delivered as a broadcast wave or OMF file.
- A mix delivered on videotape or DAT.
- A mix done in the offline delivered as a collection of audio media.

If the mix is delivered as a file or on tape:

1. Import or capture the audio mix.
2. Duplicate the online MOS sequence and move the duplicate to the Online Final bin.
3. Rename the sequence "Online Final" to indicate that it is the complete and final version.
4. Load the final sequence into the Record monitor and the audio mix into the Source monitor.
5. Sync up the audio and the sequence by the timecode.
6. The timecode should match exactly. If it doesn't, hopefully a sync pop has been provided for line-up purposes. If this is not the case, contact the offline editor.
7. Remove any extraneous In or Out marks from the sequence.
8. Edit the audio into the final MOS sequence.

If the mix was delivered to you as a collection of audio media, the media will usually arrive on a drive that you should have copied previously to a drive on the online system.

1. Duplicate the online MOS sequence and move the duplicate to the Online Final bin.
2. Rename the sequence "Online Final" to indicate that it is the complete and final version.
3. Load the final sequence into the Record monitor.
4. Open the Offline Version bin and load the offline sequence into the Source monitor.
5. Select "Clip Color > Offline" from the Timeline Fast menu.
6. Press the Toggle Source/Record button in the timeline to view the source timeline.
7. Confirm that no audio media are offline. Offline media will display as red segments in the timeline.
8. If media are offline, select the offline edit in the bin.
9. Select "Clip > Relink" to open the Relink dialog box.
10. Select "Relink offline nonmaster clips to any online items" and deselect "Relink only to media from the current project."
11. Select "All Available Drives" from the media pop-up menu and click OK.
12. Check the timeline to confirm that all media are back online. If not, contact the offline editor.

13. Park at the head of both sequences.
14. Deselect all video tracks in the online sequence (in the Record monitor) and select all audio tracks in the offline sequence (in the Source monitor).
15. Overwrite the audio into the online sequence. The two sequences should match perfectly.

Linking to Other Sequences

Sometimes you will have more than one version of the program to conform. Though it is possible that each sequence will have some unique media, it is highly likely that the majority of the media are shared between the two.

Once you have captured the media for one version you can use the Relink command to associate the captured media with the other sequence version(s).

Relinking by Resolution

One risk of relinking is that the Avid system by default will relink to the first media it finds, regardless of resolution. As you may have some of the offline media available, you can specify the resolution to relink to.

Relinking by resolution also allows you to easily exchange projects back and forth between the online suite and the offline suite, or between multiple workstations in collaborative situations where additional editing, effects work, or audio mixing are performed. You can keep two or more sets of media available in any supported resolution, and relink the sequences and clips to either.

The bottom of the Relink dialog box provides a set of options that lets you control how the system relinks:

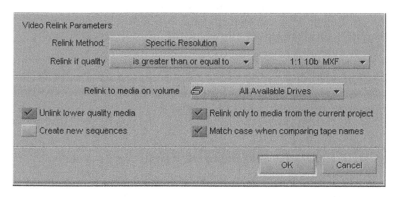

- *Relink method:* Allows you to specify how to identify the media to relink. The following options are available:
  - *Most Recent:* Relinks to the most recently created clip. This option is selected by default.

- *Highest Quality:* Relinks to the highest-quality resolution.
- *Most Compressed:* Relinks to the lowest-quality resolution.
- *Specific Resolution:* Relinks to clips of a specific resolution and enables additional options.
- *Relink if quality:* If Specific Resolution is selected as the relink method, two additional pop-up menus are enabled:
  - *Resolution:* Allows you to specify the resolution to relink to.
  - *Relink Options:* Allows you to instruct it to relink only to the selected resolution or to relink to media of the selected resolution and either higher or lower in quality. This option is usually set to *is equal to.*
- *Unlink lower-quality media:* If Specific Resolution is selected as the relink method, this option is enabled. If selected, any media that are not relinked to the specified resolution will be unlinked. This option is normally selected when relinking to alternate versions.

To relink to an alternative version:

1. Select the alternate version sequence(s) in the bin.
2. Select "Bin > Relink."
3. Select "Relink offline nonmaster clips to any online items."
4. Select "Specific Resolution" from the Relink Method pop-up menu.
5. Select *is equal to* from the Relink options pop-up menu.
6. Select the resolution you captured (usually 1:1 or 1:1 10b) from the Resolution pop-up menu.
7. Select "Unlink lower-quality media."
8. Click OK. The Avid system relinks the alternate version sequence(s) to the recaptured media. If these versions used other clips you will need to recapture them.
9. Select "Clip > Decompose."
10. Select "Offline media only and Digitized clips."
11. Click OK to decompose the sequence.

If any other clips were used in the alternate versions, these clips will be displayed in the bin for recapture. You can now recapture them from the source tapes, as required.

Conforming to High Definition

If you are conforming in high-definition video, the majority of the conform process is identical to that described above. But if the offline was done in NTSC, PAL, or even in another HD format, you must first convert the sequence to an HD format before decomposing it.

The conversion process is similar for all HD formats, but has some slight variations. Before we discuss the additional conforming steps required, let's look at some fundamental parts of the process.

Compatible or Convertible Formats

In the Avid system all SD and HD project formats are either *compatible* or *convertible* with other project formats.

- Compatible formats share a common *frame* rate.
- Convertible formats have a frame rate that is different from that used in the offline.

Compatible formats conform very similarly to SD onlines, while convertible formats require a few additional steps.

Compatible Formats: Switching a Project's Format Type

All SD projects can be switched to one or more HD formats and back again. Depending on the original project type those formats can be either 1080-line or 720-line. The only requirement is that all of the formats share an HD format and back again. (The same, however, cannot be said for 720-line HD projects.) To switch a project's format:

- Open the Project window and select the Format tab.
- Choose the desired format from the Project Type pop-up menu.

Table 9.5 lists the HD project formats that are available for each SD project format.

High-Definition and Standard-Definition Clips

Notice that in Table 9.5, the compatible projects have the same frame (or field) rate. Sequences and clips also have formats and share the same compatibility as projects. When you create a new sequence or clip (by import, log, capture, decompose, or some other method) it inherits the project format.

If an SD and HD format are compatible, then sequences and clips of the same two formats are also compatible. This means that, for example, you can open a 30i sequence in a 1080i/59.94 project and that a 1080i/50 sequence can contain 1080i/50 clips, 1080p/25 clips, 720p/25 clips, 25i PAL clips, and 25p PAL clips! Likewise, film-rate projects (24 or 23.976 fps) can also mix SD and HD material even though neither NTSC or PAL were

Table 9.5 Compatible Formats

SD Project Format	Compatible Formats
30i NTSC	1080i/59.94
	1080p/29.97*
	720p/29.97
23.976p NTSC	1080p/23.976
	720p/23.976
24p NTSC	1080p/24
25i PAL	1080i/50
	1080p/25
	720p/25
	25p PAL
25p PAL	1080i/50
	1080p/25
	720p/25
	25i PAL
24p PAL	1080p/24

*At the time of publication, the 1080p/29.97 is only available if an Avid Mojo DX or Avid Nitris DX is connected to your editing system.

originally progressive. Because the original SD material has been converted internally to progressive clips, the clips can be mixed with the HD progressive clips.

Convertible Formats: Changing a Sequence's Format

If the frame rate of a clip or sequence is not compatible with the current project format, then the clip or sequence cannot be loaded into the Source or Record monitor. You encounter this most typically at this time with the 720p/50 and 720p/59.94 project types.

In these instances you must convert the sequence's format so that it is compatible, which is done via the Modify command. Table 9.6 lists the sequence conversions that are available.

To convert a sequence, you *must* be in a project of the format you wish to convert to. To change a sequence's format:

It is not possible to directly modify the format of a master clip or subclip.

1. Create and open a project with the final online format.
2. Copy a bin containing the sequence you wish to convert into the project and open the bin.
3. Select the sequence in the bin and choose "Clip > Modify."

Table 9.6 Convertible Formats

Project Format	Convertible Format
30i NTSC	720p/59.94
25i PAL	720p/50

4. Choose "Set Format" from the modification type pop-up menu. The project's format will appear in the pop-up menu. In most instances, this is the only available option.

5. Choose the desired format from the Format pop-up menu.

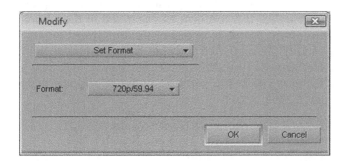

6. Click OK to change the sequence's format. In most cases, a duplicate version of the sequence will be created and the sequence name will be appended with the new format.

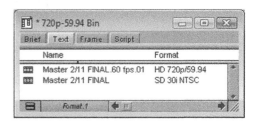

Conforming Standard-Definition Offlines to High Definition

Now let's look at the possible conform workflows and list the conforming steps required for each. The following section lists every possible SD to HD conform available in Media Composer 3.0. You can use this list as a reference when it comes time to do your next conform.

1080i/59.94 Online from a 30i NTSC offline

1. Create a 1080i/59.94 online project.

2. Copy the bin containing the NTSC offline sequence and prepare for the conform as discussed in the "Preparing to Recapture" section on page 236.

3. Confirm that the project format is set to 1080i/59.94.

4. Decompose the offline NTSC sequence.

5. Continue with the conforming workflow as discussed on page 239.

If you are using a version prior to 3.0 and are conforming from an interlaced to a progressive format, the steps listed below will be different from what is required on your system. Refer to your system's online help for the steps required in your version of Media Composer or Symphony.

1080p/29.97 Online from a 30i NTSC offline

1. Create a 1080p/29.97 online project.
2. Copy the bin containing the NTSC offline sequence and prepare for the conform as discussed in the "Preparing to Recapture" section on page 236.
3. Confirm that the project format is set to 1080p/29.97.
4. Decompose the offline NTSC sequence.
5. Continue with the conforming workflow as discussed on page 239.

720p/59.94 Online from a 30i NTSC offline

1. Create a 720p/59.94 online project.
2. Copy the bin containing the NTSC offline sequence and prepare for the conform as discussed in the "Preparing to Recapture" section on page 236.
3. Modify the sequence and change its format to 720p/59.94.
4. Decompose the offline NTSC sequence.
5. Continue with the conforming workflow as discussed on page 239.

720p/29.97 Online from a 30i NTSC offline

1. Create a 720p/29.97 online project.
2. Copy the bin containing the NTSC offline sequence and prepare for the conform as discussed in the "Preparing to Recapture" section on page 236.
3. Confirm that the project format is set to 720p/29.97.
4. Decompose the offline NTSC sequence.
5. Continue with the conforming workflow as discussed on page 239.

1080i/50 Online from a 25i PAL offline

1. Create a 1080i/50 online project.
2. Copy the bin containing the PAL offline sequence and prepare for the conform as discussed in the "Preparing to Recapture" section on page 236.
3. Confirm that the project format is set to 1080i/50.
4. Decompose the offline PAL sequence.
5. Continue with the conforming workflow as discussed on page 239.

1080p/25 Online from a 25i PAL offline

1. Create a 1080p/25 online project.
2. Copy the bin containing the PAL offline sequence and prepare for the conform as discussed in the "Preparing to Recapture" section on page 236.
3. Confirm that the project format is set to 1080p/25.
4. Decompose the offline PAL sequence.

5. Continue with the conforming workflow as discussed on page 239.

720p/50 Online from a 25i PAL offline

1. Create a 720p/50 online project.

2. Copy the bin containing the PAL offline sequence and prepare for the conform as discussed in the "Preparing to Recapture" section on page 236.

3. Confirm that the project format is set to 720p/50.

4. Decompose the offline PAL sequence.

5. Continue with the conforming workflow as discussed on page 239.

720p/25 Online from a 25i PAL offline

1. Create a 720p/25 online project.

2. Copy the bin containing the PAL offline sequence and prepare for the conform as discussed in the "Preparing to Recapture" section on page 236.

3. Confirm that the project format is set to 720p/25.

4. Decompose the offline PAL sequence.

5. Continue with the conforming workflow as discussed on page 239.

1080p/24 Online from a 24p NTSC or 24p PAL offline

1. Create a 1080p/24 online project.

2. Copy the bin containing the offline sequence and prepare for the conform as discussed in the "Preparing to Recapture" section on page 236.

3. Confirm that the project format is set to 1080p/24.

4. Modify the sequence and change its format to 1080p/24. (Even though this is a compatible format with NTSC, you must modify the format before decomposing.)

5. Decompose the offline sequence.

6. Continue with the conforming workflow as discussed on page 239.

1080p/23.976 Online from a 23.976p NTSC offline

1. Create a 1080p/23.976 online project.

2. Copy the bin containing the offline sequence and prepare for the conform as discussed in the "Preparing to Recapture" section on page 236.

3. Confirm that the project format is set to 1080p/23.976.

4. Modify the sequence and change its format to 1080p/23.976. (Even though this is a compatible format with NTSC, you must modify the format before decomposing.)

5. Decompose the offline sequence.

6. Continue with the conforming workflow as discussed on page 239.

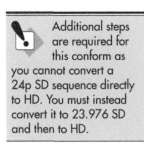

Additional steps are required for this conform as you cannot convert a 24p SD sequence directly to HD. You must instead convert it to 23.976 SD and then to HD.

1080p/23.976 Online from a 24p NTSC offline

This procedure is required if the offline was edited at the "film" rate of 24 frames per second (fps). Current broadcast delivery requirements require that all film-based programs be delivered at a video rate of 23.976 fps, which matches the slowdown applied to NTSC (29.97 fps versus 30 fps).

1. Create and open a 1080p/23.976 online project.
2. Copy the bin containing the offline sequence and prepare for the conform as discussed in the "Preparing to Recapture" section on page 236.
3. Change the project format to 23.976p NTSC.
4. Modify the sequence and change its format to 23.976p NTSC.
5. Change the project format to 1080p/23.976.
6. Modify the sequence and change its format to 1080p/23.976. (Though not explicitly required, this step is recommended, especially if you are working on a version prior to 3.0.)
7. Decompose the offline sequence.
8. Continue with the conforming workflow as discussed on page 239.

720p/23.976 Online from a 23.976p NTSC offline

1. Create a 720p/23.976 online project.
2. Copy the bin containing the offline sequence and prepare for the conform as discussed in the "Preparing to Recapture" section on page 236.
3. Confirm that the project format is set to 720p/23.976.
4. Decompose the offline sequence.
5. Continue with the conforming workflow as discussed on page 239.

720p/23.976 Online from a 24p NTSC offline

This procedure is required if the offline was edited at the "film" rate of 24 fps. Current broadcast delivery requirements require that all film-based programs be delivered at a video rate of 23.976 fps, which matches the slowdown applied to NTSC (29.97 fps versus 30 fps).

Additional steps are required for this conform as you cannot convert a 24p SD sequence directly to HD. You must instead convert it to 23.976 SD and then to HD.

1. Create and open a 720p/23.976 online project.
2. Copy the bin containing the offline sequence and prepare for the conform as discussed in the "Preparing to Recapture" section on page 236.
3. Change the project format to 23.976p NTSC.
4. Modify the sequence and change its format to 23.976p NTSC.

5. Change the project format to 720p/23.976.
6. Modify the sequence and change its format to 720p/23.976. (Though not explicitly required, this step is recommended, especially if you are working on a version prior to 3.0.)
7. Decompose the offline sequence.
8. Continue with the conforming workflow as discussed on page 239.

Conforming Video-Rate Offlines to a 23.976p Online

When conforming a video-rate offline to a 23.976p online, you have to make a series of assumptions. The first one is that the 2:3 pulldown was done so that the A frame is on either a "0" or a "5." This is usually the case for modern HD down-convert transfers. If, however, this was not the case, you will need to modify the Pull-in column and enter the appropriate timecode to the pull-down phase relationship.

Due to the nature of a "matchback" conform, the final sequence will be accurate to ±1 frame at each cut. In addition, as the sequence must be converted to an EDL as an interim step, most of the effects done in the offline will be lost.

1080p/23.976 Online from a 30i NTSC offline

This procedure is required if a production shot on film at 24 or 23.976 fps was offline edited within an NTSC 30i project. As the HD conform must be done at 23.976p, you must convert the 30i video sequence back to its original film rate before you can conform with the HD master tapes.

Part 1: Create a Matchback EDL
1. Create a new 30i NTSC project with Matchback enabled.
2. Open this new project, copy the bin containing your offline sequence into it, and open this bin.
3. Choose "Bin > Set Bin Display," enable "Show reference clips," and click OK. *This step is critical.* If you do not display the reference clips, the system will not be able to convert the timecode format.
4. Choose "Bin > Headings," activate the Start heading, and click OK.
5. In the bin, select the Start heading and press Ctrl+D. A dialog box will appear asking you to select the column you wish to copy the information to.

If Matchback is not enabled in this new project you will be unable to continue. It is not necessary, however, for the original offline project to have had Matchback enabled.

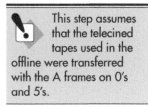

This step assumes that the telecined tapes used in the offline were transferred with the A frames on 0's and 5's.

6. Select the TC 24 column and click OK. Twenty-four frame timecode is generated for each master clip used in the sequence.

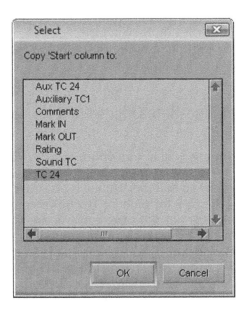

7. Load the offline sequence into the Record monitor.
8. Choose "Output > EDL" to launch EDL Manager.
9. Choose 24 from both the Source timecode and Record timecode pop-up menus.
10. Set the EDL Type to "CMX_3600."
11. Select "V1" from the video track pop-up menu. (If the offline sequence has more than one video track you need to generate separate EDLs for each video track.) Select the long dash "(for no track)" from each of the audio track pop-ups. (As only the video will be recaptured in the online, the audio source tapes do not need to be considered.)

This EDL format may result in some source reel name changes. (But other options may result in even more changes.) You will need to check the source list for a cross-reference list of sources.

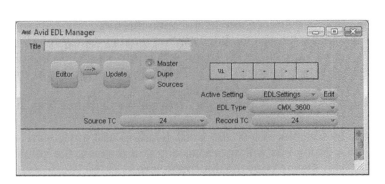

12. Click the → button to load the sequence into EDL Manager.
13. Choose "23.976" from the Project Type dialog box.
14. Choose "File > Save As."
15. Name the EDL and save it to an appropriate location on your system.
16. Quit EDL Manager. *This step is critical.* You must quit and relaunch EDL Manager for the conversion process to work. You cannot simply "round trip" the EDL.

Part 2: Create a 23.976p Sequence

1. Return to the Avid system and close the 30i NTSC project.
2. Create and open a 1080p/23.976 online project.
3. Change the project format to 23.976p NTSC.
4. Create and name a new bin that will be used for the converted offline sequence.
5. Choose "Output > EDL" to launch EDL Manager.
6. From EDL Manager, choose "File > Open."
7. Locate the EDL you saved previously and open it. A dialog box appears asking you to define the frame rate of the EDL.
8. Choose "24" and click OK.
9. Click the ← button to send the EDL to the Avid editor. A dialog box appears asking you to select the project type.
10. Choose "23.976p NTSC."

11. Return to the Avid editor. A dialog box appears prompting you to select a bin to receive the EDL.
12. Choose the appropriate bin, then name and save the converted sequence.
13. Prepare for the conform as discussed in the "Preparing to Recapture" section on page 236.
14. Change the project format to 1080p/23.976.
15. Decompose the offline sequence.
16. Continue with the conforming workflow as discussed on page 239.

> If you generated more than one EDL (one for each video track), load the EDLs one by one and generate sequences for them. Then merge the sequences together.

Conforming Mixed-Format Sequences

The Decompose function does not know how to differentiate between SD-format clips and differing HD-format clips. Instead, it will create new clips that match the current project format.

Therefore, you will need to decompose multiple times, once for one set of clips (SD or HD) and then again for the remaining clips. The order (SD then HD or HD then SD) is not significant.

The following section will illustrate the procedure for a mixed 30i NTSC and 1080i/59.94 conform. The other conforms that support mixed format clips follow a similar procedure.

1. Create a 1080i/59.94 online project.
2. Copy the bin containing the NTSC offline sequence and prepare for the conform as discussed in the "Preparing to Recapture" section on page 236.
3. Change the project format to 30i NTSC.
4. Decompose the offline NTSC sequence.
5. Sort the bin by tape name and delete all the clips that came from an HD source tape.
6. Continue with the conforming workflow as discussed on page 239 to recapture the SD clips. After you have finished capturing the SD clips, return to this workflow to continue for the HD clips.
7. Change the project format to 1080i/59.94.
8. Decompose the sequence. Be sure to select the "Offline media only" option or you will decompose the clips you just captured as well as those that came from an HD source tape.
9. Continue with the conforming workflow as discussed on page 239 to capture the HD clips and continue the conform.

10

COLOR CORRECTION

"In visual perception a color is almost never seen as it really is—as it physically is. This fact makes color the most relative medium in art."

—Josef Albers

One of the most sophisticated toolsets within the Avid editor is the color-correction toolset. Within color correction you have access to a wide range of controls and adjustments that let you make fine adjustments to the "look" of the images in your program. Avid Media Composer and Avid NewsCutter share a basic version of this toolset while Avid Symphony contains a more advanced version. In this chapter we'll look at both sets of tools but focus our attention primarily on the version in Media Composer and NewsCutter.

Before You Correct

Color correction is a very precise mix of art and science. And *where* you correct has as much affect on the image as *how* you correct. Remember that although the camera's eye does not lie, the human eye (integrated with the brain) does "lie" all the time. Our eyes correct for environmental lighting, such as incandescent lighting, and show us what our brain perceives to be balanced colors. This is why, for example, a sheet of white paper appears to be white to our eyes in incandescent lighting even though it is actually quite yellow in that light. As you can imagine, this can make color correction quite difficult! If you can't trust your eyes, how can you correct? In order to properly color correct our environment must be conducive to proper color recognition. Optimally this means your environment should have:

- A properly calibrated broadcast monitor.
- Lighting of the correct color temperature.

- A neutral gray wall behind the monitor.
- An external waveform/vectorscope.

Let's discuss each one of these elements.

Properly Calibrated Broadcast Monitor

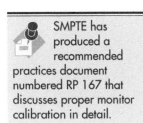

SMPTE has produced a recommended practices document numbered RP 167 that discusses proper monitor calibration in detail.

At a bare minimum you should have a monitor that has been properly calibrated to SMPTE bars so that its setup, gain, chroma, and, for composite monitors, hue are properly adjusted. But for a professional correction (or grading) suite, you should also ensure that the monitor has a proper color temperature and that it has a neutral grayscale.

The recommended color temperature for a critical viewing monitor is 6500 Kelvin or what is often referred to as D65. Many consumer monitors are much bluer in tone than a D65 monitor and will not provide you with an accurate image. For example, if the flesh tones appear too blue, but it is entirely due to the monitor, removing the blue cast caused by the monitor will result in yellow flesh tones in your images. As yellow flesh tones can be interpreted as jaundiced, you probably won't have a happy client!

Purchasing a proper viewing monitor used to be a relatively trivial matter as all manufacturers produced tube-based monitors that properly conformed to SMPTE spec. But now that the industry has transitioned to LCD (liquid crystal display) and plasma monitors, it is much more difficult to get a proper grading monitor. Though these monitors have come very close in color accuracy—especially LCD monitors that use LED (light emitting diode) backlighting—they still have a problem displaying clean accurate blacks. Significant improvements in LCD monitor quality and accuracy have been made, however, in the past two years and hopefully these problems will eventually be resolved.

It is also important that a monitor have a neutral grayscale. This means that the "color" of gray does not vary from black to white. Many consumer television sets shift toward green in the blacks and toward blue in the whites. If the monitor you are correcting on introduces a shift that you then try to remove you will create an undesirable result!

Proper Lighting

Remember that the color of an object is affected by the light that illuminates it. It is critical that the light in your room is of the same D65 color temperature as your monitor. With the advent of compact fluorescent lights you can now inexpensively order D65-based lighting. Certainly your big-box hardware store won't carry these, but a specialty electrical supply house should either stock them or be able to order them for you.

Once you have the proper color temperature you should ensure that the lighting does not directly reflect off your monitor. This will affect how you view the colors displayed and will unquestionably affect the accurate display of black. Low-wattage bulbs are recommended and will help with the reflection problem.

Proper Wall Color

Here is where you'll have to tell the interior decorator to take a hike. You don't want a richly colored wall behind your monitor! Instead you want a neutral gray wall behind your monitor. Ideally this wall should be softly illuminated with D65 lighting. This type of lighting is referred to as a *bias* light and is designed to create a halo of properly neutral gray around your monitor. Your eyes use this neutral background to keep themselves properly calibrated. Believe me, if you look at an incorrectly calibrated image in isolation it will eventually look correct to your eyes. The bias lighting helps keep that from happening.

You can take a gray card to a paint store and have them mix you a can of gray paint, but make sure that you specify that they must use a pure-white base. Many white bases are actually either warm or cool in temperature as they provide a better base for the paints consumers typically prefer.

External Waveform/Vectorscope

If you are serious about color correction, you need scopes. Period. No argument or debate allowed. When you are color correcting you aren't just making nice-looking images, you're also making images that meet a broadcast delivery specification. This means that your blacks can be no lower than 0-mV digital and your whites typically no brighter than 700-mV digital. There is no way you can do this with just your eyes. You need scopes to tell you the voltages in your image. Of course, you also need to know how to read them! Teaching you how to use external waveform/vectorscopes is beyond the boundaries of this book, but you can find excellent books on scope usage. You should also sit with a colorist and learn from him how they use scopes while grading.

Avid includes a set of internal scopes in the color corrector but these are rudimentary at best and no substitute for real scopes. Certainly they are a good tool to learn and experiment with but if you are going to make money color correcting you need real external scopes. Internal scopes are never a proper substitute.

Client Discussions

Finally, before you begin correcting it is a good idea to have a discussion with the director and/or producer of the production you'll be color correcting. You need to know about the conditions of the shoot and whether any footage was shot in a specific way, for example, with a tobacco filter on the lens. There are few things worse than assuming a special treatment is actually an error and correcting it out. That green fluorescent look might actually be intentional! You should speak with them and determine the following:

- Were there any problems in the shoot, such as white balance or lighting problems? Did anything go wrong that would affect the color and exposure?
- Was any specific filtration used or deliberate attempts made to create a specific look in the camera?
- Were any shots intentionally shot low contrast, day-for-night, or a similar specific treatment?
- Do any of the scenes in the sequence require specific looks?
- Is there any color in a shot that must be matched to a physical sample or color?

As you have this discussion, go through the sequence to be corrected with them and make careful notes as to their direction. Ask them questions about what you see and whether they want it corrected. (For example, on one episode of a program I corrected, the female protagonist's recent hair coloring reacted poorly to the set lighting and, as a result, had a slight green tint. They appreciated my noticing and agreed that it should be corrected back to natural blonde.) By taking the time to go through the sequence prior to beginning your correction, you are not only showing a level of professionalism, but are expressing a genuine concern for the quality of your client's work.

The Color-Correction Interface

When you enter color-correction mode you are presented with a special configuration of the Composer window and the Color Correction tool. Let's take a look closer look at each one.

Composer Window

The Composer window switches to a three-monitor view, each of which is configurable. By default they are configured, left to right, as Previous clip, Current clip, and a Reference image. Each monitor can be individually configured to show other items such as the entire sequence, the next shot, or one of the built-in scope

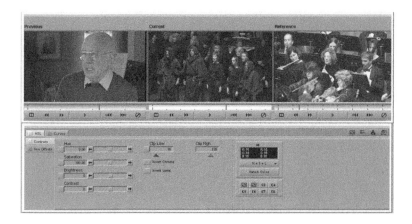

displays. If you want to change one of the monitors, simply click on the name of the monitor to display the monitor source menu. The default configuration is generally recommended and gives you two different comparison clips with the current clip you are correcting.

At the bottom of each monitor there are a series of nonmodifiable buttons. Most of these should be familiar to you from other parts of the application. The Play button has a special function in color-correction mode. Instead of playing the sequence, it only plays to the end of the current clip. The only exception to this rule is if you play within a monitor configured to either show the Reference image or the entire sequence. In these two instances it plays from the position indicator to the end of the sequence. As we'll see, though, there are other ways to play through edits, and I don't recommend that you play within the Reference configuration. The entire sequence option is fine for screening a scene or an entire sequence but there are other play options as we'll discover later in this chapter.

Using the Dual Split in Color Correction

The Dual Split button, located below each monitor in the Composer window, can be used just as in Effect mode to perform a split screen to show both corrected and uncorrected views of the current frame. Instead of using it as a split screen, though, I recommend using it as a *full-screen* switch. If you drag the lower right corner of the split display to the lower right corner of the monitor, you can use this button to quickly switch between the corrected and uncorrected version of a frame.

This is especially useful when viewing your correction in a video monitor and is a very helpful tool in preventing your-self from being "lost" in a correction. Sometimes when correcting you can lose track of the objective and what the shot started as. Rather than disabling all the corrections in the Color Correction tool, simply toggle this button back and forth and the image will quickly switch between the corrected and uncorrected versions. (I map this button to the Escape key on my system when correcting as I use it so often in a grading session.) Unfortunately, you cannot play the uncorrected version, as displayed by the dual split. If you need to play, you must disable the correction in the Color Correction tool.

The Color Correction Tool

Just below the Composer window is the Color Correction tool. This is similar to the Effect Editor in that it contains all of the parameters you will use to perform the correction. It is divided into tabs, for major correction groups, and subtabs, for different sets of parameters within a given correction group. In Media Composer, subtabs are only used in the HSL group, though they are used for additional groups in Avid Symphony, as we'll see later.

Each tab, subtab, and parameter has an Enable button that allows you to either toggle the correction on and off or, by holding the Alt/Option key, to reset the correction tab, subtab, or parameter to its default value.

On the far left of the Color Correction tool are the Match Color tool, which we'll discuss later in this chapter, and the correction buckets. The buckets provide you with up to eight savable corrections. (Versions of Media Composer prior to version 2.8 only had four buckets.)

To save a correction to a bucket, simply Alt/Option click on a bucket. Everything you have done in the active correction is saved to the bucket for later use. To use a bucket's correction, simply click on it. The buckets' corrections are stored until you quit the program. If you want to save them for another session, simply drag each bucket to a bin to save the correction. The corrections will be automatically named for the bucket number to facilitate their reloading in the next session. Reloading a bucket is as easy as simply dragging the saved correction to the desired bucket.

Let's take a look at the two correction groups available in the Media Composer (or NewsCutter) Color Correction tool.

HSL Group

This group is called the "HSL" group because it operates in the Hue, Saturation, and Luminance color space. The HSL group is divided into two subtabs, Controls and Hue Offsets.

Hue Offsets

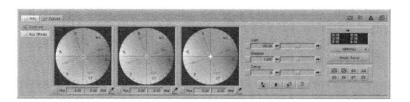

Hue Offsets is the default subtab displayed and contains the majority of the tools you'll use for primary HSL-based corrections. To the right are your three basic Luma controls, Setup, Gamma and Gain, and to the left are your three color-balance wheels. Let's look closer at these controls.

- *Setup:* Increases or decreases the luma voltage. Setup is an additive control that increases all voltages by an equal amount. It is primarily used to set the voltage position of the blacks in your image.
- *Gamma:* Adjusts the midpoint of the luma range. Increases in gamma will increase the voltage of all luma values but have the most affect on the middle voltages (those at 350 mV). Voltages near 0 and 700 mV are only minimally affected. This control is primarily used to adjust the grayscale balance.
- *Gain:* Multiplies the luma voltage. As gain is a multiplier, the higher the voltage (and therefore the brighter the value), the greater the increase. It is primarily used to set the peak white in the image, though, as a multiplier, it will affect all voltages.
- *Shadow chroma wheel:* Adjusts the white balance of the shadow region of the image. The shadow region is defined by a shelf where voltages lower than approximately 30 mV are completely affected by the adjustment. The curve then slopes downward until just past 350 mV, resulting in a gradual reduction in affect of the chroma wheel. This wheel is primarily used to balance the blacks and shadow detail in the image.

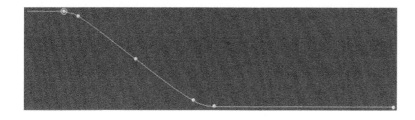

- *Midtone chroma wheel:* Adjusts the white balance of the midtone region of the image. The midtone region is defined by a bell curve with the primary affect at 350 mV, gradually sloping off in both directions until there is no affect at approximately 100 mV and 600 mV. This wheel is primarily used to balance the skin tones in the image.

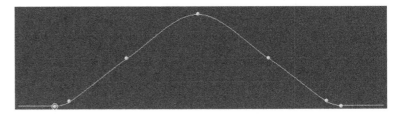

- *Highlight chroma wheel:* Adjusts the white balance of the highlight region of the image. The highlight region is defined by a shelf where voltages higher than approximately 670 mV are completely affected by the adjustment. The curve then slopes downward until just below 350 mV, resulting in a gradual reduction in affect of the chroma wheel. This wheel is primarily used to balance the peak whites in the image. Unfortunately, many images contain highlights that are actually in the midtone range, so you may need to use that control to manipulate them as well.

You can redefine the shadow, midtone, and highlight regions in Avid Symphony.

Controls

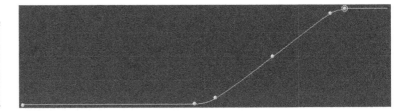

The Controls group contains six parameters and two switches that can be used for additional correction within the HSL group:

- *Hue:* Rotates the hues in the image around the color wheel. A value of +120 results in red shifting toward green, while a value of −120 results in red shifting toward blue.

- *Saturation:* Increases or decreases the color saturation of the image. A value of 100 is the default. This is the most commonly used correction within the Controls subtab.
- *Brightness:* Increases or decreases the luma voltage. Setup is an additive control that increases all voltages by an equal amount. It is primarily used to set the voltage position of the blacks in your image and is functionally similar to the Setup parameter in the Hue Offsets subtab.
- *Contrast:* Increases or decreases the amount of contrast in the image by applying a multiplier to values above 350 mV and a divider to values below 350 mV. Though this can be used to quickly establish a shadow and highlight in some images, you are typically better off using the separate Setup and Gain parameters in the Hue Offsets subtab.
- *Clip Low:* Establishes a value, or voltage, below which all voltages are hard-clipped to the desired value. In Media Composer and NewsCutter this parameter is hardwired as 8-bit values and a value of 16 is equivalent to 0 mV.
- *Clip High:* Establishes a value, or voltage, above which all voltages are hard-clipped to the desired value. In Media Composer and NewsCutter this parameter is hardwired as 8-bit values and 235 is equivalent to 700 mV.
- *Invert Chroma:* Equivalent to rotating hue by 180 degrees.
- *Invert Luma:* Inverts all voltages within the image around 350 mV.

The two Clip parameters are typically used at the end of a correction to remove any stray voltages, such as specular highlights, that were not brought into the valid broadcast range by adjustments of Setup and Gain.

Curves

The Curves group operates in the RGB (red, green, blue) color space and allows you to control the voltages of the individual R, G, or B channels separately (via their own curves) or as a group (via the Master curve).

A curve is a graph that maps the relationship of the input voltages of the image on the horizontal axis to the output values of the correction on the vertical axis. (The input is what is being fed into the correction and the output is the adjustment made by the correction.) The default lower-left to upper-right diagonal represents a neutral curve where no change is being applied to the color channel or channels. Control points are provided at the ends of the curve and you can add additional curve points to make changes to the desired channel. Points to the left of the curve are increases in voltage while points to the right of the curve are decreases in voltage.

Color-Correction Workflow

Now that we've taken a look at the various parts of the Color Correction tool, let's examine how to best use them. Color correction truly is an art and a skill that can't be learned in a single chapter or a single session. Like editing, it takes years of practice and usage for you to gain the eye and the aesthetic required. There are also excellent books on color correction that I strongly recommend, especially *The Art and Technique of Digital Color Correction* by Steve Hullfish (2008), a book that examines this art from the perspective of professionals in the industry. Another great resource is Avid's own class on color correction, which is available directly from Avid and at their Avid Authorized Education Centers worldwide.

There are many different approaches to correcting a shot but all approaches have some of the same fundamentals. Therefore, even though we can't learn the art and the craft in a chapter, we can discover a basic workflow that can be used as a starting point in your exploration of color correction.

In order to correct a shot, the following must be done at some point during the correction:

- Set a proper, color-neutral black point.
- Establish a well-balanced grayscale.
- Create a neutral highlight.
- Balance the flesh tones.

Certainly there will be exceptions to the rule—and shots where a neutral tone is not desired—but for general correction all four criteria listed above are generally met. Let's take a look at how to perform a correction that meets these criteria.

Establishing the Grayscale

The first stage to our workflow is to get the grayscale, or luma, right. Remember that in video the foundation of our signal is the luma. Therefore, it is critical that we have a well-balanced and appropriately represented grayscale before we attempt to do anything with color. If you ignore this step you'll find it very difficult to get good-looking results and will often end up with muddy results instead.

Establishing a proper grayscale typically makes use of three parameters in the HSL group: Setup, Gain, and Gamma.

Setup: Setting the Black Point

The foundation of your image is its blacks. Virtually every image should have a proper black somewhere in it. Certainly there are exceptions to this rule, but as a general rule the absence of a rich black will make the user think that the image is washed out. If this is your intention then go for it, but if not, then get that black set.

A proper black should be close to but not drop below 0 mV. (As we saw in Chapter 5, if you are measuring an NTSC analog signal you should not drop below 7.5 IRE.) Use Setup to make this adjustment, raising or lowering the voltage of the entire video signal until you have a proper black. If the shot doesn't have a pure black, don't force it to 0 mV—this is where the eye and the aesthetic come in—but use your judgment to find the correct voltage for the blackest blacks.

Gain: Establishing the White Point

Once you have a properly positioned black, use the Gain slider to adjust the overall amplitude of the signal so that you have a proper white as well. This does not automatically mean a white that rises up to 700 mV, but it definitely means a white that does not exceed 700 mV! Many shots will not have a pure white in them, even when they are properly exposed. Here again, you must use your aesthetic to find the right voltage for your white. Unless the day is overcast, exterior shots often have at least specular highlights that should approach 700 mV even if the whitest white on the subject doesn't begin to approach that voltage. This is where your scope can be a real aid to you. (You do have scopes, right?)

Leaving specular highlights far above 700 mV is recommended only if you enjoy having your tapes rejected and doing freebie make-do work for your clients. If you'd rather be paid to do your work then don't leave them up there! But don't just use the Luma Clip parameter to hard clip your whites. This will definitely be necessary in some cases, especially for interior shots with blown-out windows, but you should try to get the most image you can within the 0- to 700-mV range before applying your clips.

For example, on a bright day the clouds can contain specular highlights well above 700 mV. If you hard clip those clouds at 700 mV you'll end up with white shapes that lack depth or definition instead of proper clouds. If you're convinced you have a good reason to do so, remember what many great teachers say to their art students: Before you can break the rules, you must *know*

To help you focus on the grayscale, you may want to temporarily set the chroma saturation to zero. After you are happy with the look of your black-and-white image you can set it back to 100 and adjust as required.

You've probably noticed that I'm talking about art and aesthetic in this chapter. I'm not doing it just so I can channel my college art professors; I'm doing so because I firmly believe that great colorists paint with light just as great cinematographers do. And, a poor colorist can destroy the brilliance of the best cinematographer as easily as the best colorist can remove the imperfections from an image and make it truly brilliant.

the rules with every fiber of your being. In doing so, you create art. Without such knowledge, though, you create randomness.

After establishing your white, you may discover that your black has changed its position. Because gain is an amplifier, all voltages are affected. Always check your blacks after making a gain adjustment and correct as necessary. Then make a quick check again of your whites and tweak if required. You'll find a lot of interactivity between these two adjustments, so be prepared to go back and correct any new problems that arise.

Gamma: Balancing the Grayscale

Once you are satisfied with the black and white of your image, take a look at the overall grayscale. Does it appear to be well balanced or is it weighted too heavily toward the blacks or whites? Certainly some shots are supposed to be dark, but don't let the grayscale be crushed to black. The Gamma adjustment changes the grayscale balance and lets you properly position the majority of the signal. If you would like your shot to be lighter in tone, increase the gamma. If you'd like it darker in tone, decrease the gamma. Just remember that you are making an adjustment that can affect the entire image, black to white. Don't forget to check your black and white points after adjusting gamma as they may well have been affected, especially if you had to move them by a significant amount originally.

The following illustration, reproduced in color in the four-color insert, shows the before and after grayscale adjustment of a typical interior image (shot on HDV with available lighting and a blown-out window). Two additional shots are also provided in the insert. Note how not only does the corrected shot have a better balanced grayscale, but the voltages in the waveform monitor are also better balanced. (See Figures 1, 2, and 3 in the color insert for color examples.)

Correcting for Color Balance

Once you are satisfied with the grayscale it is ready to move onto the chroma. Just as you adjusted your blacks and whites first before working on the overall grayscale, the classic approach to color balance is to make sure that your blacks and whites are pure and color neutral. This is especially critical for the blacks though arguably less so for the whites.

Balancing the Blacks

As mentioned previously, the blacks are the foundation of your image and they must be properly positioned, both in luma voltage and in color purity. You want to make sure that your blacks are *pure* black and not polluted with any color cast. Though you think you can do this with your eyes, this really is

Original Image Prior to Correction

Corrected Grayscale Image

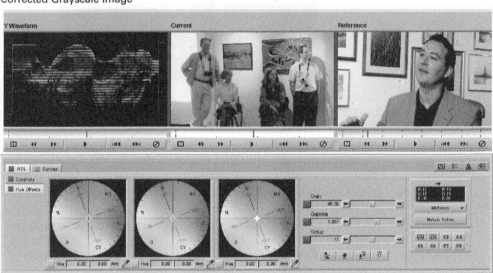

best accomplished using external video scopes. If you have an analog scope, the best pattern to use is the vectorscope with the I line indicated. On a digital scope you can use either the vectorscope and I line, or the RGB parade.

Balancing the Blacks with a Vectorscope

A vectorscope is a display that shows you only the chroma in the image. The available chroma is displayed in a polar graph with the various hues arrayed around the circle like the hands on a clock. As saturation increases, the display plots the hues further out from the center of the graph. A lack of any chroma is plotted in the center.

As we want our blacks to be perfectly neutral, we need to identify our pure blacks and center them on the graph. Though this is easiest done with a real-time external scope, we can use the internal vectorscope with some practice. (The primary disadvantage of the internal scope is that it is not "live" and interactive, but instead only updates when you release the mouse button.) To identify the blackest

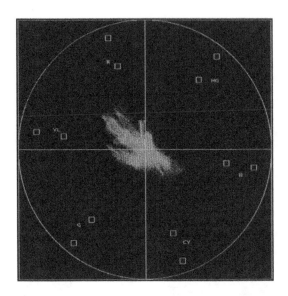

blacks, grab the control cross in the Shadow wheel and drag it around (releasing it every so often if using the internal vectorscope). As you drag you should see in the vectorscope a small black mass or, less technically, blob that responds directly to your movement. That blob represents your shadow tones, and if you center it in the middle of the vectorscope you neutralize them, removing any visible color cast. (Even if you are ultimately going to shift the color balance of the image to something other than neutral—for example, to provide a cool blue look often associated with a technical facility—you still want properly neutral blacks in almost every instance.)

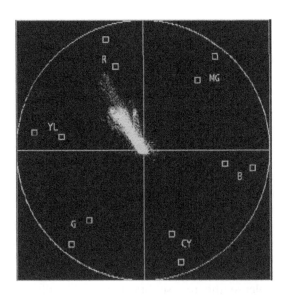

Balancing the Blacks with an RGB Parade Display

An RGB parade is a waveform display where the voltages of calculated red, green, and blue components are displayed side by side, left to right. These are typically calculated values because, as you'll recall from Chapter 5, a video signal is comprised of luma and two color-difference signals, not R, G, and B. Because the signal is ultimately decoded back to R, G, and B, though, this type of display is extremely useful. Unfortunately, due to the nature of its creation, it is typically only found on digital scopes, and not analog.

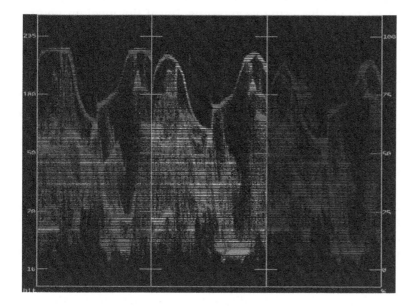

As with using the vectorscope, the goal is to neutralize the blacks. In this case, our blackest blacks are the lowest points on our three parade graphs. To neutralize the blacks is to align the lowest voltages so that they are equivalent. To accomplish this you use the Shadow wheel and drag it (releasing it every so often if using the internal RGB parade display) until the three patterns are aligned at the bottom. As you are using two axes to control three results, this takes a bit of practice to accomplish! You should practice this technique, though, because experienced colorists should be able to do this as if it were second nature.

Balancing the Whites

Now that you've balanced your blacks, you can move onto the whites. If there is a distinct bright white in your image and it *should* be white (such as, for example, a sheet of paper or a pure-white wall) then you can use either the vectorscope or RGB parade display similarly to the approach you took with the blacks. But if there isn't a distinct region this is a bit trickier to accomplish. Indeed, if there isn't a distinct black some colorists tell me they actually approach the flesh tones before the whites.

Adjusting for Flesh Tones

The midtones are typically where we want to make the greatest adjustment and, if there are faces in the image, the flesh tones reside here. If you do not have any people in the shot, see if you can identify a neutral object in the midtones and use the same technique you used for balancing the blacks. It will probably be easier to do this using vectorscope, but if the medium neutral object is large and distinctive enough, you may be able to identify it on the RGB parade display and use it instead.

If, however, there are flesh tones in the image, you are better off balancing the midtones by these instead of a neutral object. This is because the human eye is trained from the earliest age to identify with faces and will notice an incorrect flesh tone long before it identifies anything else as being incorrect.

For some, balancing the flesh tones is one of the trickiest parts to correction. How do you know whether a flesh tone is right or not? Fortunately, the logic behind how a video signal is stored and captured can help you. On an analog vectorscope you'll typically see two lines etched into the graticule. These lines, known as "I" and "Q," are key vectors in composite analog signals. They are also axes that happen to correspond to flesh tones and green foliage. The I line corresponds to flesh tones, and because all humans share a common blood, the foundation of our skin color

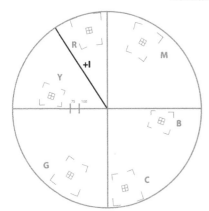

is essentially the same. Though there certainly can be exceptions, the reality is that nearly all human flesh tones reside somewhere on or near the I line. Unfortunately, this line is not displayed on the built-in vectorscope, but it is readily available on analog and digital vectorscopes.

To balance flesh tones with the I line, simply drag the midtone wheel back and forth until you identify the region, or blob, of the signal that represents the flesh tones. Then, move that region until it is on or near the I line on the vectorscope. You can then "walk the I" up and down, along the angle of the line, to either warm or cool the flesh tones. It really is an amazing technique to know as a colorist.

Color Correcting with Curves

This group is called the Curves group because it allows you to redefine the output of the calculated Red, Green, and Blue channels using an *xy* graph. The Curves group contains four graphs: one each for red, green, and blue, and one called Master, which represents red + green + blue.

You also see four parameters that are duplicated from the HSL group: Master Setup, Master Gamma, Master Gain, and Master Saturation. These allow you to make either luminance adjustments (with Setup, Gamma, and Gain) or saturation adjustments. It is important to understand, though, that these "Master" adjustments are processed prior to curves and the results of these changes are fed into curves. This means that you can, for example, change the saturation of an image that you will then adjust with curves, but you cannot remove saturation from an adjustment made in curves. This order of processing is critical to the color-correction process and cannot be changed.

The Curves group will be familiar to you if you've worked with Adobe Photoshop. This is an incredibly powerful tool because it allows you to remap each channel in a linear or nonlinear fashion. Indeed, when you first display the Curves group you won't see anything that resembles a curve, but instead see four diagonal lines. That is because by default the input values (on the *x*-axis) and the output values (on the *y*-axis) are equivalent and there is no change to the image. But if you add a point and remap input to output, a curve appears that describes the change from input to output.

Understanding Curves

The Curves group makes it easy to understand what the general result of an adjustment to a curve will be by the background colors displayed for each curve. Let's take the Red channel's curve as an example. If you add a point (or adjust one of the two initial points) and drag it up and to the left you will add red to the image. If you drag a point down and to the right you will add cyan to the image. Why cyan? Well, what you're really doing is removing red from the image and, according to color theory, when you remove a color you can be seen as adding that color's complement. The complement of red is cyan, just as the complement of green is magenta and of blue is yellow.

The Master curve does not have any color cast because it applies its adjustment equally to all three color channels. You might think of this as a luma adjustment, but in reality it isn't because it affects both the luma and chroma signals. This is because, as we recall from Chapter 5, both luma and chroma are used to calculate red, green, and blue.

Though my personal preference is to do the majority of my correction work with the HSL group, the Curves group is extremely powerful. I usually use this group to apply color treatments or casts, rather than for general correction. That said, I know many colorists who can very quickly do general correction with Curves. It really comes down to your level of comfort and experience with the tools.

Beyond color treatments or casts, I find that the Curves group can be used for signal clipping, especially when you are trying to remove chroma overvoltages that are difficult to otherwise remove. If you hold down the Shift key while dragging the upper control point, the point will travel diagonally down the existing curve, adding a clipping line behind it. I use this technique all the time when I have, for example, overly hot yellows that are overdriving my composite signal. A simple adjustment—for example, to the Blue and Green channels—are often all it takes to correct an otherwise problematic yellow and make it suitable for broadcast.

You will recall these magic numbers from both Chapters 5 and 7. The range 16–235 corresponds to 8-bit values for the voltages from 0–700 mV.

You can also use this technique to precisely clip the R, G, and B signals at 0 and 700 mV, which will also limit your luma to 700 mV, by utilizing specific values for the input and output axes. Below each curve are two data entry fields that correspond, from left to right, to the input and output axes. Setting both input and output to 16 results in a clip at 0-mV digital, while setting them both to 235 results in a clip at 700 mV.

If you want to take this clip a bit further you can also add two additional control points between the two and drag the higher one up and to the right and the lower down and to the left to create a classic S-curve shape. This shape has the effect of stretching the voltages in the signal, resulting in a greater amount of contrast. If you want the opposite result and a reduction in contrast simply move the two points in the opposite direction. I use the contrast-reducing S-curve less frequently but have found it to be extremely useful for treating backgrounds in a graphics composite.

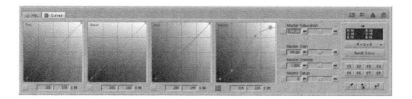

Correcting White Balance Problems

Another useful technique for Curves is to remove white balance problems in images. This is one of the most common problems I see with video productions because either the shooter forgets to white balance, or especially in documentary work, there just isn't time to do a proper white balance when the characters are moving quickly from an exterior to an interior. Take a look at the following illustration, also reproduced in the color insert, Color Figure 4.

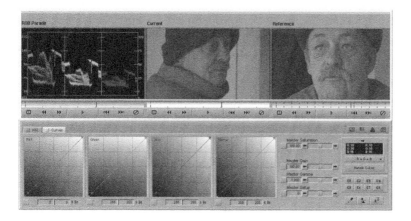

As we can see, most specifically in the color version of this illustration, the current shot is far too red, especially when compared to our properly balanced reference shot. The RGB parade also illustrates this lack of balance as the red channel is dramatically higher than either the green or blue. By lowering the red channel while making a slight boost to the blue we are able to remove the majority of the incorrect color cast. You'll notice from the color illustrations that we left a bit of the warmth in the shot as compared to the reference as the producer preferred a slightly warm look for the scene.

As with the grayscale adjustments, we've provided two additional white balance corrections in the color insert (see Figures 4, 5, and 6 in the color insert).

Creating Color Treatments

Finally, the Curves group is especially useful for creating color treatments. For example, we have to be "trained" by watching to feel that warmer colors equal a warmer actual temperature and perhaps a happier scene, while cooler colors equal a cooler actual temperature and perhaps a sadder scene. And certain locations have a distinctive look. Los Angeles is typically portrayed as warmer, while New York is typically portrayed as cooler. (For a classic example of this type of treatment, compare the overall looks of *CSI: Las Vegas*, *CSI: New York*, and *CSI: Miami*.)

Let's look at how to do a couple of typical color treatments. In each instance you should refer to the color insert for the examples as they are difficult, if not impossible, to represent on the black-and-white page.

For our first example, we'll take a relatively warm scene and cool it down. The producer wanted to represent a sadder time in the character's life and we've chosen to cool down the shot. Cooling down a shot is generally accomplished by reducing both red and green in varying amounts (usually more reduction in red though some shots will require more reduction in green) and increasing blue slightly. In this instance, we also lowered the overall luminance of the shot as well by reducing the Master curve in the midtones (see Figure 7 in the color insert).

In our second example, the producer wanted a desaturated, mildly sepia look for a flashback section. This was accomplished by using the Master curve to reduce the overall contrast, lowering the Master saturation, then slightly increasing the Red curve in the midtones, while lowering the Blue curve in the midtones (see Figure 8 in the color insert).

Playing within Color Correction

As we mentioned earlier, the Play button has a special function within color-correction mode. Instead of playing through the sequence, it will only play a single segment in the timeline. This is designed to make it easy to analyze a shot prior to correction and to play the correction after you've made it.

But there are times when you need to play through an edit without leaving color-correction mode. In these instances, Avid has provided special functionality while in color-correction mode to two other play buttons.

- *Edit Review:* This button works in conjunction with the Trim Preroll and Postroll settings to play a portion of the surrounding clips. (Remember that the Trim Preroll and Postroll settings are located in the Play Loop tab of the Trim settings.) For example, if the Trim Preroll and Postroll are set to a default of two seconds, the Edit Review button will play the last two seconds of the previous clip, then the current clip, and finally the first two seconds of the next clip.
- *Play Loop:* This button instructs the system to play the entire sequence in the active monitor, starting from the position indicator's current location. This button is extremely useful if you want to play through a section of the sequence without switching monitors or exiting color-correction mode.

I use Play Loop so extensively in my color correction work that I've mapped it to the Tilde key.

Color Correcting with Avid Symphony

Avid Symphony contains a superset of color-correction capabilities that exceed those of both Media Composer and News-Cutter. It also has different ways of representing a correction. Color corrections can appear in the timeline in one of three ways:

- A green line on the bottom of the timeline, which indicates that a color correction has been applied at the source level.
- A blue line at the top of the timeline, which indicates that a color correction has been applied at the program level.
- A color-correction effect, which indicates that a color correction has been applied as an effect. These are most typically found when a color correction was originally created on Media Composer, but they can also be created within Symphony.

Source and program corrections are perhaps the most powerful aspect of Symphony's color corrector as they allow a single correction to be applied across multiple segments in the timeline. For example, a source relationship can be tape based and automatically applied to all shots from a given tape. This is especially useful if the entire tape was improperly white balanced. A single correction could fix all of the shots used from that tape. They can also be applied to all segments edited from a single master clip. Program relationships are track based and allow you

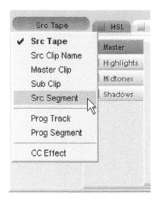

to apply a single correction to all shots on a given video track or even, by using marks, to a range of clips on a video track.

You can select a relationship by choosing one from the Relationship menu, located at the upper left hand corner of the Symphony Color Correction tool. The source and program relationships are processed in parallel. If you select the CC Effect relationship you can add a serial cascaded correction.

Now let's take a look at the additional controls and groups available within Avid Symphony.

Additional HSL Controls

The HSL group has been dramatically expanded and includes a number of additional parameters, tabs, and subtabs. Let's take a look at each tab and discuss the differences therein.

Controls

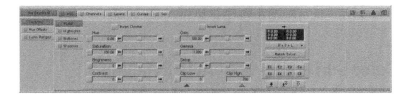

There are a number of significant differences from the Media Composer color-correction toolset.

- *Parameter location:* All of the slider-based parameters are contained in this tab, instead of being divided between the Controls and Hue Offsets subtabs. This means that you should look here for your Setup, Gamma, and Gain parameters.
- *Highlights subtab:* Provides a full set of parameters—with the exception of the Invert and Clip parameters—that only operate on the brightest regions of the signal. These parameters are especially useful for manipulating blown-out regions of the frame. You can bring the blown-out portions down to manageable levels without bringing the entire image down in gain.
- *Midtones subtab:* Provides a full set of parameters—again with the exception of the Invert and Clip parameters—that only operate on the midtone region of the signal. These parameters are perhaps the least used of the HSL controls, but can be quite helpful when you need to work more closely on the flesh tones in an image. The Midtones Gamma parameter is especially useful in these instances.

- *Shadows subtab:* Provides a full set of parameters—with the exception of the Invert and Clip parameters—that only operate on the darkest regions of the signal. These parameters are especially useful for expanding—or stretching—the shadow range. For example, if you are satisfied with the overall look of the image but want deeper, darker shadows, you can use the Shadows Setup parameter to stretch the shadows downward.

Arguably many of the parameters are overkill. For example, you might wonder when one might use the Shadow Highlight or the Midtone Setup, especially as they control very similar regions of the signal. But the engineers decided to err on the side of completeness rather than remove a portion of the controls only to have someone need them.

Hue Offsets

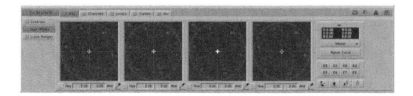

This tab is the least changed in the HSL group. The color in the wheels has been removed to provide a basic vectorscope-like display. You can reenable the color, though, from the Color Correction settings. In addition, a fourth, Master parameter wheel is provided. This wheel is used less frequently than the other three, but can be used to make an overall change to the image. If the shot has a severe color cast, you may be able to remove most of the cast with a single Master wheel adjustment. Use it with discretion, though, as extreme adjustments of this parameter can cause the signal to exceed typical broadcast delivery limits.

Luma Ranges

This tab is used to customize the luma ranges that define the highlight, midtone, and shadow areas. Any customized ranges will be used for both the subtabs within the Controls tab and the wheels in the Hue Offsets subtab. Each range can be individually adjusted.

If I make an adjustment to these ranges, I typically open up the highlight range. As mentioned previously the default configuration of this curve only provides full affect between 670 and 700 mV. This is a fairly narrow range and I often see what I would define as a highlight extending down to nearly 600 mV. By reshaping the curve I can include these voltages in my adjustments.

Channels

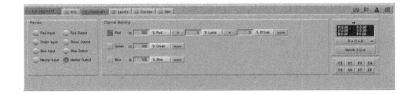

The Channels group is best thought of as a channel repair tool. In it you have the ability to modify the signal by adjusting how the RGB primaries are generated. This group is especially useful if, due to poor white balance or a damaged encoder chip, one of the Red, Green, or Blue primary channels in the image is either very low in voltage or clipped. You may be able to recreate this damaged channel by blending in information from the other channels, from Luma or one of the color-difference signals, or apply an offset. In some cases, especially if there was in-camera clipping on one of the color primaries, these controls may mean the difference between a usable and an unusable shot.

You can add or subtract up to 200 percent of any given signal then apply a positive or negative offset to the end result. Up to four signals can be mixed together to create the new color channel. To add a new component to the formula for a channel, simply click the More button to the right of the existing components.

Levels

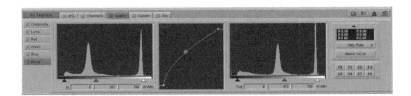

The main function of the Levels group is to balance the contrast range, or tonal range, of an image. You do this by defining the white point, gray point, and black point of the video material. The group provides a great deal of control over the amount of contrast and detail that is visible in the video image, especially since you can make adjustments on the individual color channels. Think of the Levels group as a more refined and specialized version of the Setup, Gamma, and Gain controls in the HSL tab.

In essence, the Levels group is very similar to the Curves group in that you are adjusting input and output points for a correction process. The major difference is that it can operate not just on the RGB channels, but also on the luma and the entire composite signal. Both of these options only have black and white point controls and not gray point controls. One approach to using Levels is to use it to set precise positions for your luma black and white, rather than relying on the Setup and Gain parameters in the HSL group.

Secondaries

The Secondaries group is the most powerful group within the Symphony color correction. All other parameter groups are primary corrections that affect the overall color balance of an image. A secondary correction, on the other hand, only affects a specific range of color within an image. Symphony can define and correct up to 12 simultaneous color ranges.

Secondary color correction works best when the range of colors that you want to affect does not appear elsewhere in the image. For example, you can use a secondary color correction to change the look of the sky as long as the blue tones of the sky are not present elsewhere in the image.

You define a secondary vector by specifying a range of hue and saturation. You can then modify those colors' hue, saturation, and luminance. For example, you might want to correct a shot of a person wearing a bright yellow scarf. After earlier stages of color correction, you are pleased with the image in general, but the scarf remains too saturated. Applying a secondary color correction to the yellows allows you to bring down the saturation of the yellows in the scarf without affecting any other colors in the image.

The Symphony Secondaries provide two different types of vectors: standard and custom. Standard vectors allow you to select a range of hue and specify a minimum saturation. They look like pie wedges. Custom vectors allow you to select a range of both hue and saturation and look like ellipses with the defined hue and saturation ranges used as the two axes of the ellipse. You define them by specifying a midpoint for both hue and saturation then a range for each.

Conclusion

As you can see, the color corrector in both Media Composer and Symphony is incredibly sophisticated and powerful. A full exploration of the color corrector is worth its own book, but hopefully we've provided you with a primer that can get you started with these powerful tools.

11

TROUBLESHOOTING

"Trouble will come soon enough, and when he does come, receive him as pleasantly as possible. Like the tax-collector, he is a disagreeable chap to have in one's house, but the more amiably you greet him the sooner he will go away."

—Artemus Ward

Troubleshooting is a detective story, an Easter egg hunt, and a test of method, memory, and patience. It is rarely an excuse for panic, since it is only by calmly tracing problems to their roots that you will ever achieve the solution. The end result is to get back up and running to finish the job and complete the vision. You cannot separate the medium from the tools, even if you prefer to think of Avid as a cloud of magic that exists at the far end of the cables.

Basic Troubleshooting Philosophy

You will never know how a system works until you know how it breaks. This means you need to understand signal flow and basic connections. When a part of your toolset is missing you need to know where it comes from and be familiar with the points of failure. For instance, if you are trying to capture video, but receive nothing but black, you need to work your way backward to each stage that might be wrong. Is the Capture toolset right? Is the video source type correct? Is the FireWire cable to the computer connected? Is the Avid hardware turned on? Is the correct video cable connected from the deck to the Avid hardware? Is it connected to the outputs of the deck? Is the cable bad or not fully connected? What is the signal supposed to be on the tape? You get the idea: Trace backward to isolate the single point of failure in the signal flow.

As you form the hypothesis of the area that should be investigated (i.e., video flow from the deck), be sure to check only one area of suspicion at a time. This means you have two approaches depending on the nature of the problem. In the case of a loss of video, you could use a subtractive procedure of checking one area before moving on to the next, such as replacing a problematic cable or trying another setting (like trying different video monitor inputs). Eliminate a potential failure point, and move on to the next.

Or you may want a more additive approach where you disconnect everything attached to the computer and add them back one at a time. This additive method would apply to problems with drives, networks, and other peripherals.

It is also useful if you can't boot the system at all. Attach a peripheral and reboot. If successful, then add another peripheral and reboot again. If changing something doesn't fix the problem, then go back to the original state before you move on—there is no need to complicate the search by eating the breadcrumbs back to where you started.

Keep track of every piece of software that is added to the system and when it was added. If someone has added software that loaded a driver (or another innocent piece of secret software) and now the Avid software won't launch, you should know what to uninstall. Advanced troubleshooters should know how to get into the Windows registry to delete recalcitrant ".dll" files, but for now just be knowledgeable about uninstalling software and launching to the last-known good configuration.

There are actually some very easy ways to get back to work without much effort. First, if you encounter an error message, but you're allowed to continue, then quickly save and keep going. If there is odd behavior after the error, you may be better off either quitting and relaunching the application or rebooting the machine. If you are on Windows then check the event viewer. Right-click on "My Computer" and choose "Manage," then "Event Viewer." You will see if there have been numerous errors in a particular area that may focus your troubleshooting.

RTFM

There are many everyday situations where just a little knowledge of troubleshooting can keep you going forward, give you a bit more confidence, and, maybe, help you keep your job. The first thing that can help you instantly is to read the Release Notes. This falls under the category of RTFM (*r*ead *t*he *f*ine *m*anual), but some people skip it because they just want to be up and running with the newest version no matter what. Stop, smell the roses, and read the known bugs. Avid is pretty good about listing what they consider to be the bugs you need to know. Your definition of a bug and their definition of a bug may differ, but you will definitely benefit by seeing that, for instance, one small part of something you need to do all the time doesn't work under certain conditions. It also helps to know if that procedure has been replaced by something faster, better, and simpler. Even if you haven't memorized the Release Notes, they can generate a little thought bubble over your head if something seems familiar. So don't toss the Release Notes; keep them handy and even scan them quickly before you call Avid customer support. Over time the base of Avid users has migrated to a larger percentage using Windows-based systems and now the balance is about 50/50 with the Apple systems. Current Avid systems are shipping on OS X and Windows XP, but there are plenty of users still on Windows NT, Windows 2000, and Macintosh OS 9. x and earlier. For all of these versions there are still some universal basics for troubleshooting.

Customer support has improved significantly over Avid's early days of explosive growth into a new industry and Avid takes support very seriously, but for you, that's not really the point. There is time involved in figuring out that you have a problem, realizing you don't know how to fix it, telling the other people in the room that maybe they had better get some coffee, and then dialing for help. It is better to say something like, "Hmmm, did you know you are missing the active SCSI terminator?" and subtly imply, "Aren't you glad you hired me?" However friendly, competent, knowledgeable, and good-looking Avid support may be, you want to avoid talking to them until you have a serious problem.

Techniques for Isolating Hardware from Software

Part of the trick of troubleshooting is to determine, before you go very far, whether the problem originates with the hardware or the software. If you have just installed new hardware (like RAM, or replaced an old board with a new board) suspect that first. Image distortion, with Meridien or ABVB systems, was typically hardware failing instead of software (though of course there is always the exception). No hardware can function unless it has the correct software to interface with the operating system. So it might not be the hardware itself that has a problem, but the driver for the hardware. This can be a device driver, and Avid uses .sys files for device drivers in Windows. For OS X, Avid uses kernel extensions (even though they are not technically drivers, for the most part they are the same as a driver). These drivers should be installed in standard updater software as soon as the hardware is in place, but if you are replacing a board you may have to load a new driver. Hopefully, this little bit of software came with the new board and will install itself with a simple double-click and a restart. Don't pull out a "defective" new board until you have installed any new software updates that may have been included.

Hardware Problems

There are four major areas of hardware problems with Avid systems: PCI boards, drives, monitors, and keyboards/mice. All these areas include cables and connectors that go bad, and they can all go bad if yanked hard enough. You can isolate drive problems by shutting down or disconnecting the drives. You can isolate monitor problems by switching the cables to another monitor. Unless you have spilled sugary brown liquids on your keyboard or mouse, then cables and drivers are guilty until proven innocent.

Software Problems

There are two major areas where you could have problems in the Avid application: the project files, bins, and user settings, or the individual media itself. Try to separate problems with the media playback from problems with the individual media itself. Media can always be recaptured, but I'll bet you don't have a few extra video cards hanging around. The best way to tell if a particular problem rests with a piece of media is to take a close look at it. Step through the problem areas frame by frame. That will show if the problem is there when the drives are not working so hard. Any corrupt images or crazy colors that are visible when you are looking at a still frame can be solved by recapturing that shot. If you can't see the problem, you must go one step deeper.

Is the problem playback related? See if the sequence plays back without any of the media online. Go to the Macintosh Finder level and dismount your drives. On the Macintosh you can do this by dragging the drives to the trash (the drive icons, not the folders inside). On all systems you can change the name of the folder on the media drives from OMFI MediaFiles or Avid MediaFiles to anything else. I just add an "x" to the beginning of the name so I don't get into trouble trying to spell the correct name of the folder again when I am in a hurry. As discussed previously, any change to the name of the OMFI MediaFiles folder causes all the media inside to go offline. Now does the sequence play? If there are no errors, you know that the problem is related to the media, the drive, or the SCSI/fiberboard. If you continue to get obscure errors (my favorite is BadMagic), then you need to look at the media databases, sequence, or the actual media.

Media databases are small files in every OMFI MediaFiles folder. As mentioned in Chapter 4, currently they are called "msmFMID" and "msmMMOB" on Windows and "msmOMFI. mdb" and "msmMac.pmr" on Macintosh. Sometimes these files become corrupted and do not update correctly. This may result in files not appearing in the Media Tool or media appearing offline when you know for sure the media are there. You must force the media databases to be updated and you can do this by forcing the update with the pulldown function Refresh Media Directories. If this doesn't work you may be forced to delete or move the media database files to another folder. This will force a recreation of the file from scratch. Moving or deleting the media database file will force a rescan of all of your media drives, which may take a few minutes, depending on how much media you have. After the rescan and recreating of the media database files, you may have successfully herded all of your lost media back online.

If the sequence or the media are the problem then you must practice "divide and conquer" to find the offending media file or element in the sequence. If you suspect that it is bad media, not just the drive, then bring all the media files back online before using this technique. Then split the sequence in half and try to play the first half. If it plays, then try to play the second half. If you have problems with the second half then divide that in two and repeat the procedure. Continue this until you isolate the area of the sequence that is corrupt. It may be a single graphic, master clip, or effect. Delete the effect, replace the graphic, and either extract or overwrite the master clip. Consider deleting related precomputes and re-rendering. Recapture the master clip to see if it is just that media. If you do not have access to the original tapes or files, consolidate the media file to another drive and see if it plays. If the problem remains, you may have to relog and capture the master clip to eliminate the problem.

There may be something corrupt in your bins, your project, or your user settings. This is so easy to test that many times it is one of the first things an Avid support representative asks you to do. Generally, I suspect this particular problem when something that has been working fine all day stops working or features that should be available are suddenly not there. Create a new bin or project and drag the clips and the sequence you were working on into the new bin.

The next step is to remake your user settings. This should be a fast check to see if creating new default user settings will clear up the problem. If it does, then spend the time to recreate your settings. Better yet, call up an archived version of your user settings that has been hidden where no one can get to it. You should always have a backup of your user settings, and as a freelancer you will want to carry them with you. These user settings may also be corrupted, but chances are pretty good that they are not. Again, make sure you have new user settings for every major change in the software. If you call up a really old group of user settings, it may be incompatible with new menus and functions in subtle, but important ways.

If the user settings are causing the problem, delete the old settings from the User Settings folder. Do not throw away the AvidDefaultPrefs in the User Settings folder since this file holds all the standard bin headings and will not be recreated automatically. Make sure you know what files are in the Site settings before you delete those; the files may contain vital standards settings that are required for every project. You can cover yourself by moving these files into a temporary folder rather than deleting them. If moving them does not solve the problem, move them back.

Audio Problems

In general, if you have an audio playback problem check the audio meters, the sample rate, then the cabling. You may solve the problem by powering down the computer and cycling the power on the Avid audio hardware. Check obvious issues like mixer power or speaker power. Trace the audio flow through the whole system, making sure you have a signal at the beginning, and go step by step to the end.

Quickly check audio meters by using Ctrl/Command+1 to call up the Audio Tool and playing the media. If there are no levels at all then the audio media may be offline or at a different sample rate than the Audio Project setting. You may want to click on the PH button in the Audio Tool and choose "Play Calibration Tone." If you can see the levels, but can't hear the tone, then you have a cabling problem with the output. Turn down the volume before you try this method—it will be greatly appreciated by everyone else in the suite.

A sample rate is the amount of audio samples that are played back per second to reproduce digital sound. There are four common sample rates in video editing today: 32 kHz, 44.1 kHz, 48 kHz, and 96 kHz. There are higher sample rates but they are used mostly for high-end digital audio workstations like ProTools HD. Sample rate mixing in ABVB systems was a problem never addressed until Meridien systems. You can't play back multiple sample rates in the same ABVB project, and you can't convert from one sample rate to the other within the Avid software. Both of these issues are solved in Meridien-based systems and later. Use SoundAppPC with Macintosh, a freeware utility that can be used to convert audio sample rates for ABVB systems, or recapture the audio at the correct sample rate.

To track either problem, isolate the sequence into its own bin, and through Set Bin Display, choose "Show Reference Clips." Choose bin headings Offline and Audio Sample Size, then sort (Ctrl/Command+E) to find which clips have audio offline and which clips are at the wrong sample rate. With a Meridien system or later you can convert the individual clips to the proper sample rate using "Change Sample Rate" under the Bin menu. Alternately, you can go to the Audio Project settings and choose "Show Mismatched Sample Rates as Different Colors" and look at the sequence. If you have different colors in the timeline then you should consider converting the sample rates or changing the Audio Project setting "Convert Sample Rates When Playing" to "Always."

You may experience clicks when editing some digital audio because the edit cuts in the middle of an audio sample. This might be especially prevalent on imported audio CDs. You can fix this quickly by adding two frame dissolves between the cuts.

Error Messages

If you are not used to working on complex professional software, you may not be used to generating error messages. Write down the ones you see and, if they do not keep you from continuing, call Avid customer support after the session and get the official explanation. Saying, "I got an error" isn't enough to let Avid customer support help you figure out the problem. It may be something systemic or it may be operator error. But if it is a "Fatal Error," you should be on the phone immediately if you don't know what caused it. After working on the system for a while, you will learn what causes the most common errors, and if you keep a log of when they occur and to whom, you can identify the pattern that applies to your own system, facility, and way of working.

The real trick to error messages is deciphering them. They may be colorful but essentially meaningless until you figure out what the function is that has gone wrong or what kind of pattern they follow. A section of the programming code always generates a specific error message when something goes wrong. Essentially, error messages tell you what happened, not why. This is because the same error could have been caused for ten different reasons. The computer cannot look outside of itself and say, "You have too many unrendered effects at 2:1, the wrong terminator on media drive 4A, and that last sound effect on audio track 6 just put me over the edge!" It will say "Audio Underrun," because what happened was that it could not continue to play every frame of audio and video through that segment of the sequence. Many times this is the only explanation that the computer can confidently produce for why it cannot play.

The Importance of Connections

Let's look at the basic world of connections. You may have enjoyed playing air guitar to Molly Hatchet, but air SCSI doesn't work as well unless it's correctly connected to something. Connections are one of the first things to look at, especially if you have just moved the system. "I just moved my monitor and now it is broken" will lead most support representatives to check your cables to the monitor.

Many of the connections to the computer have electricity running through them and connecting anything "hot" can cause a component to burn out. A good rule of thumb is not to change connections while the computer is running (not including audio or video to an external source like a deck).

Cables that are screwed in tight don't come loose so quickly. This seems obvious, but many facilities decide it is easier to have them not screwed in so anyone without a Phillips head screwdriver can move things and make changes quickly. Do yourself a favor and buy a screwdriver with one end Phillips head and the other end flat. Tighten everything on the back of the computer and anywhere else you can tighten things down. Of course, if you yank really hard, you still have problems, because now you have loosened the computer board it was attached to or even ripped the wiring out of the connector!

Make sure that things are fastened down away from big feet and spilling coffee, but don't pack them away so tightly that you can't get at them to take a look. Get a few bags of little plastic tie wraps from your favorite electronics store and wrap cables together in logical groups. You have to cut the tie wrap to pull them apart at some point, but this is still better to discourage the unauthorized, and you can always wrap them again when you are done. The cost compared to downtime is negligible.

Good engineers leave a little slack in all their carefully tie-wrapped suites so that any piece of equipment can slide forward enough to see around back. Large, easy-to-read labels that are also easy to understand make any phone call to Avid customer support less embarrassing. "Is the video input connected?" "Uh, you mean cable VI649?" If you need to call your own video engineers, at least you will be sure that the problem is not software related. Also contemplate moving everything in the Avid suite a few feet away from the wall and mounting a small, clip-on lamp back there. This gives the system some more air if the room tends to get warm and will give you easy access.

USB and PS2 Cables

One of the silliest problems that seems to happen often is with the keyboard cables. USB is the newest version, but some Windows systems also use PS2 cables, especially for keyboards. You don't want to have too many USB devices connected at once, and be careful about using USB extenders. This is a good place to look if peripherals are acting strangely. First, let me caution everyone who is thinking of pulling this harmless-looking USB cable and reconnecting it while the system is running. You may have done this a hundred times, but the next time you do it, you can fry your keyboard or even your motherboard. Any cable that has power running through it has the capability to create a power surge or damage sensitive electronic parts by "hot-swapping" or changing the connection while the power is running through it.

Unfortunately, all keyboard and peripheral computer cables are always about an inch or two too short and so are under a certain amount of tension all the time. Many times just a slight pull is all it takes and your system appears to crash. You try all the keyboard reboot commands and nothing works and finally you go to the computer processing unit (CPU) itself to press the reset switch. When everything comes back online—surprise! You still don't have control because what really happened was the keyboard cable had come just slightly loose.

SCSI Connections

A USB connection is basic and straightforward compared to the scariest of all computer connections—SCSI (small computer system interface). The term "SCSI voodoo" may not be completely foreign to you and for a good reason. Even though you may follow all of the complicated rules of SCSI or simplify your system to minimize them, you may still encounter situations that just don't make sense.

Basic SCSI Rules

There are some basic rules to keep in mind no matter what your SCSI configuration is. If you want peak performance and as many streams of real time as possible, you should consider still using SCSI drives. In many cases, fiber drives will be sufficient, but FireWire drives will not be much good for the higher resolutions and full amount of advertised real-time streams. FireWire drives get faster and larger all the time, so check the specifications to make sure you can get the throughput you need. If you are cutting long-form, single-stream, or dual-stream projects like feature films and documentaries, you may find the FireWire drives sufficient. For everything else, SCSI and fiber are recommended, and for uncompressed high definition (HD), SCSI is still the best. Of course, when we see 10-Gbps fiber switches, all bets are off!

On older Macintosh systems there was an internal SCSI card with every system. It was really meant for things other than video streams. For example, you could connect up graphics, audio and backup drives, removable media, scanners, or the ancient external three-dimensional (3D) effects "pizza box" (the Pinnacle Aladdin) to this internal SCSI connector. The real power was in the SCSI accelerator card that Avid included to connect the media drives. Most high-end PCs have an internal SCSI connector that is quite fast and may be enough for your standard-definition (SD) needs. If you have another card, however, you can add more drives to the system. And if you use a dual-port SCSI card you can get even more speed from your drives through striping (which we will cover later). Keep in mind, the more things you have connected at once, the more complicated your SCSI troubleshooting will be.

Keep the length of your SCSI connections as short as possible. The cables that ship with your drives are meant to be that short because anything longer than the maximum length of normal SCSI chain causes serious voodoo behavior. There are many supported types of SCSI drives: Classic SCSI-1, Fast and Wide SCSI-2, Ultra SCSI-3, and Ultra2 LVD (a variation of SCSI-3). Each type uses different cables and different cable lengths. So rather than try to outguess the manufacturer, stick to the length that comes with the system. The length of the cable used for the entire SCSI chain must take into account the length of cable inside each drive. That can add up pretty fast at over a foot per drive. A common mistake is to take one look at the length of the cables that came with the drives and rush out to the electronics store and get the longest cable you can find. Avoid the temptation to use these and keep the cables short.

Make sure all your SCSI cables are of the same brand, type, and style—different cables may have different internal configurations that may cause some devices to just never work right. This is why it is so important when mixing different types of drives to get all the cabling and termination correct, as we will cover later.

SCSI cables are very sensitive to twisting and must be handled more carefully than any other cable on the system. Just bending an SCSI cable back and forth a few times can significantly reduce its functionality. Don't strain, kick, stomp, or bite them.

Finally, turning drives on and off has its own set of rules. Make sure all SCSI devices are turned on and have clearly come up to speed before powering on the CPU. Listen for each drive to make a single "click" sound when it has finished spinning up. Keep all SCSI devices on if they are connected in the chain to ensure consistent behavior. Turn off all peripherals only after the computer has shut down.

There are lots of complicated rules about SCSI, especially when you are combining the newer ultra drives with the older narrow drives. These four terms—ultra, fast, wide, and narrow—refer to the speed and capabilities of passing larger amounts of data through the SCSI chain. The drives themselves don't look all that different.

Drive Striping

You can get better performance from any drive by combining it with other drives in a striped set. Drive striping is a way to make many drives act as one large, fast drive. There are five main types of drive striping, but the two to remember for Avid systems are RAID 0 and RAID 1. RAID 0 (redundant array of independent disks) is what most people mean when they stripe drives together for more real-time effects. You use the drive striping software that comes with Avid, mount all the drives, and stripe them into one large drive with multiple read/write heads and thus faster seek time to find or write media. This process will erase all the media on the drives, so do it only when they are new out of the box or when you have backed up all important material. You can even stripe drives across multiple SCSI connections if you have a dual-channel SCSI PCI board. If you have four newer drives and a dual-channel SCSI board, you can create what is called four-way striping with two drives on each SCSI channel. It is highly recommended to do this on the Adrenaline systems for a maximum of real-time effects based on many streams of video. Four-way striping is a good way to continue using older, slower drives, too, but there is a drawback. If one drive fails, you will lose all the data on the other three drives, too.

RAID 1 is also called drive mirroring and is used in Avid Unity MediaNetwork. This means that you have a duplicate drive for every drive full of media and the duplicate is created during capturing. You can decide to turn this on or off on Unity depending on your needs and the importance of the project. If you choose to use it then you may not even notice when a drive fails. You will continue to keep editing and can replace the failed drive when you have some time to spare. Avid is working on systems that fail even more elegantly, without duplicating drives. These systems will eventually heal themselves by spreading redundant data across all drives, so that when a drive fails, it can be recreated on a spare in the background without an effect on performance.

Termination

All SCSI chains must be terminated. Termination is the way the computer knows where the end of the chain is and which drive is the last device. It keeps the signal from bouncing back to confuse the computer with false signals. Technically, the chain must be terminated at the beginning and at the end, but on most CPUs (all the newer ones and all the ATTO cards), the termination at the beginning is internal. This means that you need to attach the terminator that comes with the drives to the last drive in the chain. Don't worry about termination at the beginning of the SCSI chain (although occasionally that internal termination can fail, too). Always be sure to use an active terminator, the purple terminator that came with the narrow drives or the blue one that came with the termination kit for the wide drives. With the new low-voltage differential (LVD) drives the terminator is beige and has an LED that indicates whether it is in the LVD mode (green) or single-ended (amber). We'll discuss the difference when we deal with mixing drive types. Don't use the generic gray terminators from the electronics store—they are generally not active.

SCSI ID

The most basic SCSI rule is that the SCSI ID, the number associated with this drive in the chain, must be unique to the chain. This usually becomes a problem when you add a new drive to the SCSI chain or bring a drive over from another system. Always check to see that the new drive does not have the same SCSI ID number as another drive already on the system. The SCSI ID is set on the back of standard drives with a pen or other pointed object; on MediaDocks it is set on the front panel. The fact that this fundamentally important piece of information is a small number on the back of the drive again points out the importance of having access to the equipment after it is installed. How do you know which numbers are being used unless you can stick your head back there with a flashlight and read them upside down and backwards? The best way may actually be another piece of software—the Avid Drive Utility (ADU) on the Mac or the Computer Manager on Windows XP. These utilities show you the correct SCSI ID even if the ID number on the back of the drive is broken and displaying a wrong number! ADU has a facility to flash the lights on the front of the drives, so if you physically label the drives, this can help when troubleshooting. You can also cause the drive lights to flash by copying a small file to that drive or striped set. Of course, you need to be able to boot your system with the drives attached in order to use this software. The new LVD drives finally have the SCSI ID on the front of the drive.

Getting the wrong SCSI ID may cause the system to not boot correctly or even to damage data. On a Macintosh system, the desktop may also indicate that you have many more drives connected than you really do. When there is a problem with the SCSI chain on startup with an older Macintosh, you may get the flashing question mark icon. This is because the system cannot find the startup drive with the System folder on the SCSI chain. You have somehow confused the SCSI chain that has the internal Macintosh startup drive.

With narrow drives, you can use only the numbers 0–6 for an SCSI ID. A wide drive can use 0–6 and 8–15. *Never* use SCSI ID 7 on any SCSI device since this is the number used by the SCSI card itself (which is also technically an SCSI device) or by the host computer. On the older, internal Macintosh SCSI chain, avoid SCSI ID 0 because that is the ID used by some internal system drives. Internal CD-ROMs generally are set at the factory to use ID 3, so when attaching devices to the Macintosh SCSI chain, you should avoid SCSI ID 3 as well. Scanners and Iomega Zip drives may use SCSI ID 5 or 6. When you are adding and subtracting drives from any system, the ID is the most important factor in making sure the drives work happily together.

Most new computers use the IDE (integrated drive electronics) or SATA (serial advanced technology attachment) internal drives and IDE or SATA connectors for CD-ROM, DVD-RW, and other internal devices, so you don't need to be concerned about an internal SCSI chain. However, you will still need to take into consideration that an SCSI board (either one that ships with the basic computer or an Avid-installed accelerator) will take up an SCSI ID, which usually will be 7. If you have a choice about adding peripherals other than drives, like scanners, CD-ROM burners, or anything else you want to connect to your Avid system, choose USB or FireWire over SCSI to help eliminate conflicts.

Connecting Wide and Narrow Drives

We have already discussed the difference between narrow and wide drives and, when you upgrade to wide or ultra drives, you need to figure out how to connect them to the same SCSI chain with the narrow drives. The slower (narrow or wide) drives must come last in the SCSI chain.

When connecting wide and narrow drives on the same chain, you need a special termination kit from Avid. Wide drives use a

68-pin connector and narrow drives use a 50-pin connector. The difference is more than cosmetic, however, since there are more active pins inside the 68-pin connector, and these extra pins must be terminated before connecting to the narrow drive. You get a new wide cable to go between the last wide drive in the chain and the first narrow drive, a small blue terminator/adapter that allows this cable to connect to the narrow drive, and a blue terminator to go onto the last drive in the chain.

Use the standard wide cable (68 pin to 68 pin) to go from the computer to the first wide drive. Then use the termination kit cable (also 68 pin to 68 pin) with a special blue terminator adapter (68 pin to 50 pin) added to it from the kit when you connect the wide and narrow drives together. Finally, use the special blue terminator on the last drive in the chain.

Cables to connect the wide SCSI card to the narrow drives (68 pin to 50 pin) are very different from cables that appear similar for connecting wide drives to the narrow drives (68 pin to 68 pin with a 50-pin adapter/terminator). Even though this narrow cable may connect, you will have nothing but problems connecting wide drives to narrow drives. It is a good idea to label these cables "for narrow use exclusively." It is better to lock up the 68-to-50-pin narrow cables somewhere after upgrading to wide drives.

Connecting LVD Drives

Ultra LVD are the fastest SCSI drives offered by Avid. They can easily play back 1:1 resolution with a two-way stripe across two LVD disk controllers. LVD drives require the same 68-pin cables as the fast and wide do, but if you connect an LVD drive to an SCSI chain with a slower drive, it drops its speed to match. In other words, an LVD drive that is connected with an iS Pro drive will perform only as fast as the iS Pro. With the cost of drives plummeting these days, it is a good idea to use only your fastest drives for media playback on the SCSI chain. Take the older drives and connect them to a server for backup over a network or for graphics and compression stations.

There are still many complications to the SCSI chain, but these are the basics. Even if you follow all these rules, you may find that a particular drive works on ID 4 but not on ID 2. You may never find a good reason for this (although there is a reason). You may also find that the order of the drives in the chain makes a difference even though all of the termination is correct at the end of the chain. And above all, you may have to juggle extra devices on the internal SCSI to find the best order. Slowly rebuilding the SCSI chain one device at a time and rebooting is often the only way to isolate where the problem is occurring and identify the problem device, cable, or terminator.

Audio Connections

Audio connections have multiplied in type and format over the years. There are now digital stereo pairs (AES/EBU XLR connections), S/DPIF, Optical, and analog (multipin octopus cables on the Adrenaline and Nitris and RCA connectors on Mojo). On Nitris there are two microphone inputs for live-audio punch-in for voiceovers and multiple language overdubbing. On Nitris and Adrenaline there are also quarter-inch phono jacks for monitoring so that you no longer need any splitter cables for audio outputs.

With this cornucopia of choices, you may want to simplify everything with a small, inexpensive audio mixer. This will allow you to permanently connect many audio sources like tape decks and DVD players. Also make sure that all of your digital devices are calibrated to the same reference level. Many older systems are calibrated to $-14\,dB$ and this caused a problem with inputting audio from Digital Betacam decks. See Chapter 9 for a full discussion of audio calibration.

If you have self-powered speakers, connect the speakers to a power strip with everything else and turn them on through that strip. This is a good idea for most of the equipment since it lessens the possibility that a piece of equipment will be left off accidentally. If the speakers are on and you still have no sound, check the Audio Tool (Ctrl/Commmand+1) to make sure you actually have audio playing back. Change the timeline view to show Media Offline and see if the audio segment in the sequence timeline lights up bright red. Using the Audio Project settings, make sure that the sample rate is set to the same rate the audio was captured at or set so that it will convert between the two sample rates when playing. The sample rate of the audio can be confirmed with a heading in the bin or by looking in the Console, which clearly will tell you that you have a sampling mismatch. This means you will not have to recapture just to hear the audio play but you will still need a roll of black gaffer's tape for something, so keep it handy!

The Black-Burst Generator

A black-burst generator (BB Gen) is like a synchronizing clock for video or audio. It provides a steady source of a perfect video signal: black. If you do not use a black-burst or sync generator, your audio and video sync may drift apart over time. This is especially obvious when capturing or outputting long takes or sequences with analog sources and decks. I highly recommend getting a black-burst generator if it doesn't come with the system you ordered.

Connect the BB Gen to decks, monitors, and your Avid hardware. Looping it through several devices is fine, but don't overdo it. Eventually you will attenuate the signal too much through multiple devices and it will fall below the minimum signal level to be effective. Digital decks like Sony's Digital Betacam will supply sync to Avid during capture. When you have set your Audio Project settings to accept a digital signal then the audio sync is provided by the deck. If you shut off the digital deck you will get an error message complaining that you have lost sync. You should switch the Audio Project settings back to analog sync.

Many decks require a stable signal to their composite video input when they are playing back, especially the Sony UVW-1800. If you are using the component inputs and outputs, consider connecting the BB Gen to the composite video input permanently. It makes it easier to "black" tapes when you are not using the deck for anything else. Just set the deck input to composite, set the correct timecode, and start to black and encode a tape with no other input. Just remember to switch it back to component input before you leave the room!

Connect signal generators like black-burst, color bars, and tri-level sync as permanently as you can and design a system that requires as little connecting and disconnecting as possible. Consider patch bays, mixers, MediaDocks, and networks. The less wear and tear you put on cables, the more reliable and long lasting they will be. A little more initial investment during the planning stages can positively reduce troubleshooting downtime in the future.

Standard Computer Woes

Even if you successfully eliminate the potential for problems with connections, there is still the potential for standard computer-type problems. Unfortunately, these kinds of problems are not easily solvable by the typical IT department, even if you are lucky enough to have one. This is one of the reasons it is a good idea to be able to take care of your own computer problems. Standard computer support personnel are going to be at a loss with Avid problems unless they have been through some training for the specific requirements of high-resolution video.

Access to Original Software

One of the last solutions to difficult, intermittent, or unusual problems is to reload the software. Avid customer support may ask you to reload the Avid software, the Macintosh or Windows operating system, or both. All users of the system should have access to these disks. A big mistake, commonly made, is to lock up these disks safely away from anyone who might need them, probably at 3 AM. Make sure the disks are the most recent and correct versions of both the operating system and the Avid software. You may cause even more problems by loading an unapproved version of the operating system!

Access to the Hardware

None of these techniques does you any good unless you actually have access to the computer itself. Some installations have cleverly hidden the system away in another room or rack-mounted it in a machine room. If you are completely forbidden to touch the hardware because of facility or union rules, then forget about it. You can just hand the phone to the appropriate authority. Otherwise you must, phone in hand, be able to look around back and see that all the cables are tightly connected, that the power to everything is on, and read the disks' SCSI IDs.

This means that you should look into getting a good engineer to set up your suite. Good engineers are worth much more than their salaries when they save you the embarrassment of your first several jobs going out the door with bad levels because you were monitoring the audio in the wrong place! Make sure, before the engineer leaves, that you have a thorough understanding of the cabling and get a wiring diagram you can refer to.

Media Management

Another day-to-day concern mentioned before is the management of media on the drives. You may think that you are efficiently deleting all media as you finish each project, but you may be surprised to find all sorts of odd bits of precomputes and imported graphics and so forth floating around on your media drives. The best way to keep a handle on this is to delete media from the Media Tool and not through individual bins. That way you see the project from a big-picture point of view and can evaluate on a project-by-project basis what must go and what must stay. It also keeps you from accidentally deleting media from another project just because you dragged a duplicate of someone else's master clip into the bin you are now deleting.

The real problem, not only with having many unnecessary small objects on your drives slowing down performance, is that the drives may accidentally become too full. You should always keep a minimum of 10–15 percent free space on any partition or drive. There are all sorts of files, like the media file database, which occasionally must get larger to accommodate the changing nature of the media on the drives. Don't worry about defragmenting media drives, but overfilled drives corrupt media and may crash and take everything with them.

A more difficult problem to diagnose is when you actually have too many files in a single folder. If you are working with 9-GB partitions and capturing offline resolutions, you may find that on long or complicated projects you are exceeding the limit on the number of files that can be in one folder on a Macintosh. This occurs because Avid must restrict the disk cache to the minimum figure in order to reduce digitizing underruns. This can be 96k on older systems and 128k on recent systems. This limits the practical number of files in the OMFI MediaFiles folder to around 1200. If there are more than 1200 files (or approaching 1200 depending on the size of the files), you may have difficulty booting the system or opening the OMFI MediaFiles folder. If your system is exhibiting these symptoms, you may want to raise the amount of the disk cache (in the Memory Control Panel), move items to another folder, and change the disk cache back again later. If you are always working with offline resolutions, you may want to limit the drive partition size to a maximum of four gigabytes.

Deck Control Tips

There are several other reasons why you might not have deck control. When you open the Capture Tool, you may see the message "No Driver." The first thing to do is to try to reload the deck configuration or force the system to "check decks." Both of these choices are at the bottom of the Capture Tool under the Deck Model pulldown menu. If this doesn't work, then you need to do some digging. Obviously, check the cable connections first. You may have a V-LAN or VLX, which is an external deck control device from Videomedia that gives you a wider range of deck control choices. If so, then make sure you are using the V-LAN or VLX cables and not the Avid-supplied deck control cable.

If none of this works, you may want to check your Release Notes or user manual to make sure you are using a supported deck. Even if the deck is not supported, you should be able to get some limited control using the generic deck choice in the Deck Template window of the later versions. With some DV devices, the only way to get your deck recognized may be to do an auto-configure. This will usually result in a generic device, which can then be switched for a particular template using the Deck Configuration settings.

Basic Windows Troubleshooting

The main objection I hear many editors use to avoid switching from Mac to Windows (besides the politics of "monopoly" versus "insanely great") is the fear of troubleshooting a more complicated operating system. The ease of access to the Macintosh System folder is a double-edged sword, and moving vital resources in and out on a whim is the cause for many editing problems. It is also something that many editors have spent time to learn and so can do some basic troubleshooting on their own. Switching to Windows means you must relearn the top ten things to do when a problem arises.

A serious problem with Avid Windows editing systems is when somebody changes something from the original shipping configuration and the system will not boot. This is handled easily by the Windows OS during the boot process when it gives you the opportunity to invoke the last-known good configuration. Press the space bar and then type "L" when prompted. This will bypass any changes the last user may have tried and will load a configuration that worked the last time the system booted successfully. You may then try again to get the configuration to be the way you want; however, if the present configuration did indeed boot correctly, then it will be the last-known good configuration, and you will need to look elsewhere to find the problem.

Windows XP Recovery

Windows XP is designed with a feature called System Restore. This keeps track of the system configuration and saves them at crucial times as known reference for when the system was running well. If you are having problems starting your system you will want to take the choice of starting in Safe Mode, which disables part of the operating system that may be causing problems. Then you can use System Restore, uninstall problematic programs, or run diagnostic programs on your disk drives.

If you are extra cautious about restoring your system you may want to look into a third-party program like Norton's Ghost®, which has many options for creating mirrored copies of your system drive. You would be able to switch over to the mirrored drive if you were under pressure to continue working without time to troubleshoot.

Reinstalling Avid Software

If you are missing a file or a file has become corrupted, you may want to reinstall only those problem files. By loading the

Avid software installation CD-ROM, you can do custom installations to replace drivers, codecs, or anything else you may suspect as being a problem.

You may want to reinstall the entire Avid application just to start from a clean slate. You may want to move all your AVX plug-ins and AudioSuite plug-ins to another location so they don't get erased; however, if you suspect that the plug-ins may have something to do with your problem, you may want to reinstall them as well. Be sure you have all the registration information for plug-ins that require being registered the first time you use them.

Version Numbers

You should always have, either in the back of your head or written down someplace accessible, the version number of almost everything associated with your system. First and foremost is the version number of the current Avid software. You should know it down to the last digit because every small change in the software has a reason. Avid puts out what they call the gold version of the software, the most tested and most stable version they can achieve within the time allowed before it must be released. Version 11 is an example of a gold release. Release dates are based on complicated interrelationships all lining up at the same time. If, after the gold version is shipped, some features didn't make it into the software, even though they were planned, there may be other releases with an extra decimal point. Release 2.1 is an example of that since it included the color-correction interface for Symphony. There may be some procedures that were not tested and appear to have problems, so there may be another revision called a *patch release*. A patch release is meant to fix one or two small problems. This would be version 6.5.1v2. The important thing to remember about all these releases is that you might not need any of them except the gold version. The other versions have not been tested as thoroughly as the gold release because of the importance of getting out a fix in a timely manner. Most of the time this is not a problem, but asking for a patch release if you don't absolutely need it pushes the envelope unnecessarily.

It is always best to have all the systems in your facility running the same version of the editing software. Most versions are forward compatible, but not backward. This means that a bin created in version 5.5 opens fine in version 6.5, but not the other way around. Once a bin has been opened in the higher version, it has been converted forever to that version. By just opening the bin once in version 6.5, you may not be able to open it in version 5.5 again. You have to export the bin full of clips as a shot log and import it into the earlier version. Unfortunately, this does not work well for sequences. A bin converter program goes between versions 6.5 and 5.5.1, but it is best not to depend on such complications.

If you are forced to change systems in the middle of a job, always insist on the same version of the software or later. You should know the version of the operating system you are using. Some versions are not approved for some models, and you should make sure that your computer can run the latest software before you install it.

Always make sure you know the latest version of Macintosh operating system that Avid has tested. If at all humanly possible, try to get all your Macintoshes to run the same Macintosh operating software. Even experienced editors can make their Avid system unstable by loading an unapproved copy of the operating system. Do not do this casually, and make sure someone else doesn't do it for you just to keep you up to date on the latest must-have features. Keeping everything interchangeable is a valuable goal and should not be complicated by an IT person or an especially enthusiastic editor who wants to put the newest software on the machine as soon as it is available.

Version numbers also carry over to the hardware. Each of your boards has a revision number, which should be considered during troubleshooting. You may have an old version of a board or a version with a known conflict. You can check the revision of your hardware in several places. If you have Avid software running, look under the Hardware Tool (in the Tools menu) to see configuration and drive use information. A utility called Avid System Test allows you to get more detailed information about each PCI board (if you have any). This is the better answer if you can't actually launch the Avid editing software. If Avid System Test (called the Avid System Utility on older systems) can't see the PCI board at all, calling it an unknown board, or if the slot is empty, then that tells you the problem is with the board or the way it is mounted in the slot.

Every utility that ships with your Avid system has a specific version number. The ones that ship are the ones that are meant either for that hardware or that software. Grabbing a version from another system just because it is newer may get you into trouble. Certain versions of the Avid Drive Utility, for instance, are designed primarily for four-way drive striping. The rule of thumb is: If the new version has a tangible improvement, completely compatible with your system and supported in the Release Notes, only then should you load it onto your computer.

Electrostatic Discharge

The sneakiest and hardest problem of all to diagnose is one that is very easy to create: damage related to electrostatic discharge (ESD). You may not realize it, but the human body can store and discharge frightening amounts of static electricity. Shuffling across the carpet with a relative humidity of 10 percent generates 35,000 volts! Compare this to the smaller, faster devices that are needed to achieve the kind of performance necessary to keep the system running happily, and they have a range of susceptibility of several hundred volts.

You can zap a component with static electricity by touching the outside of an ungrounded device or, more probably, by opening the device to do some simple troubleshooting. You may be asked to remove or add RAM or to reseat a troublesome board. Any time you are going to open a case, be concerned about voiding a warranty or causing ESD damage to the sensitive components inside.

The key to touching anything inside the system is to be grounded. Being grounded ensures that any buildup of static electricity is channeled off to a ground and dissipated. This is best done by wearing a grounding wrist strap and connecting it to a ground or to a metal component inside the computer. Most important, make sure the computer itself is grounded. The best way to ensure this is to plug the computer into a grounded outlet. An even better solution is to plug the computer into a grounded power strip that is plugged into the wall socket and then shut the power strip off. That way you are not supplying power to the computer while you are working on it.

The scariest thing about ESD is that it doesn't always kill—sometimes it just maims. A board or a RAM chip that receives a substantial shock may not fail right away. It may not fail for days, weeks, or months. It may start to show intermittent behavioral problems that cannot be isolated. These are absolutely the worst kinds of problems to troubleshoot because they may not occur for long periods and may not be of a type that points to any one component. It may be the CPU itself that received the shock, and no matter how many boards you replace, it does not solve the problem. This is why ESD should be treated so seriously; any time you handle a component or open the computer, you should be very, very careful.

Calling Avid Customer Support

If, after all these precautions and general maintenance, you must still call Avid customer support, at least take heart in a very short, best-in-class wait time. To make it go even faster, have certain answers prepared since almost all support calls start with the same basic questions. Know your versions, operating system, CPU model, and Avid model. Be able to describe what you were trying to do when the problem occurred. This is especially crucial for video engineers who have not taken the time to learn the software. The editor describes the problem and the engineer cannot explain it to the support representative in enough detail. What is the exact wording of the error? Some errors are pretty obscure like "missing a quiesce." Write it down and don't fake it with "something was missing, I think." When did the problem begin? Right after you put in the new RAM? Pull the RAM out. Does the problem happen every time you perform a particular operation or is it really random? Can you repeat it? The simplest answer to any error that seems random is to shut down the computer and restart.

The other very important thing you need to give to customer support is your system ID. If you don't have a system ID, you may not get any support! That is the way Avid determines whether you or your company has a valid support contract. This is not something you want to discover at 3 AM. The system ID can often be found by running a utility called either Dongle Manager or Dongle Dumper, but if you can't launch the computer, that won't help much. Write it down or use Dongle Dumper and print it out.

Consider how much more helpful your customer support representative will be if you do not immediately launch into a tirade of abuse. They know you are frustrated or you wouldn't have called. They are trained to help and to help calm you down, but you can make everyone's job easier by being civil and professional. And if you don't hear the hold music, you are not on hold, so don't say insulting comments about them to everyone else in the suite. The representative may have his or her headset muted, but he or she is still listening!

We can't deal with all the techniques, potential problems, and error messages in this short space. If you want to know more about your system, there are courses in troubleshooting and a whole curriculum to become an Avid Certified Support Representative. You don't need to be a technician to keep your system running happily most of the time, but you should have some basic knowledge of what is going on under the hood. Good maintenance routines and a healthy dose of caution are two necessary items when dealing with sophisticated and complicated systems. Keep it simple and you will be rewarded with fewer steps and less stress when you need to troubleshoot.

12

NONLINEAR VIDEO ASSISTANTS

"The best servant does his work unseen."

— Oliver Wendell Holmes

With the adoption of new technology has come the blending of postproduction responsibilities. Producers and writers become offline editors, offline editors become online editors, and the difference between film and video begins to blur. One thing that this shift has created is many new people who can edit well, but who don't have the inclination, ability, or time to get involved in the technical requirements.

This opens up the possibilities for an important position: the nonlinear assistant. There have always been film assistants whose responsibilities are pretty clear. They handle all the day-to-day requirements of film handling, organization, and preparation for the editor. There have also been video assistants in the past, although their roles have changed through the years and at times have been eliminated altogether. It is inconceivable to lack a film assistant on a major feature, but many high-end production companies operate quite well without video assistants.

Assistants are very important if the design of the postproduction facility is focused around a central machine room. The editor initiates the communication via an intercom system and tapes are changed, set up, and dubbed by a voice on the other end. Occasionally, the assistant is in the same room and can speed up the editing process by doubling as a sound engineer or a character generator operator.

The elimination or devaluing of the video assistant makes it harder for young people to break into the business. Since the film assistant on a nonlinear project may work a second shift while the editor works throughout the day, there is less opportunity for the interaction between master and apprentice. With tense clients who are paying large sums per hour for time-critical work, the production company that puts an unknown or untested quantity in the driver's seat is taking a risk. They may lose their client forever to the competition or may have to discount the session to appease them.

How, then, do you break into this business? There are no second video assistants like there are in film, so what is the entry-level position? Many times it is whatever the facility needs at the time: a tape dubber, a receptionist, even a courier; however, with the advent of nonlinear editing, there is the nonlinear assistant. In large and busy nonlinear postproduction facilities, there may be one nonlinear assistant per shift and three shifts per day. The entry-level position then becomes the graveyard shift and eventually the day or evening shift, where editors can discretely observe skills that keep them in demand. The job doesn't really require the ambition to become an editor, but the people who gain the most from the nonlinear assistant position are those who need to know these subtle skills to move on to the next level.

What responsibilities should these assistants be expected to perform? Much of the knowledge they need has been covered in this book. In fact, many editors perceive many of the techniques in this book as something only an assistant would perform. Others see it as required knowledge before starting a job! It is when a facility desires such a specialization of labor, either for personnel or billable reasons, that the assistants have the most value. They must perform all of the functions and have all the knowledge required to keep the systems running. The postproduction supervisor instructs the assistant on all the requirements to keep as many jobs running smoothly as possible. The administrator or supervisor sees the big picture, and the assistant performs the tasks.

These important daily tasks include capturing, media management, basic maintenance, backup, and output. Anything that is required to prepare the suite for the editor and the smooth transition from one project to another is appropriate for the assistant. Each facility has its own set of requirements, but mastery of all these skills can make someone very valuable to any busy postproduction facility.

Capturing

Capturing also implies following logs, creating bins, setting levels, and understanding drives. The logs are handed off with the understanding that the marked takes or possibly everything should be captured and the master clips named according to the description in the logs. Bad logs mean bad bins unless the assistant knows something about the specific job and is given the freedom to create better master clip names.

- Assess each take as to whether only video or only audio should be captured, thus maximizing use of disk space. Don't capture video for the voiceover!
- Check to see if everything, even incomplete takes, should be captured.
- Don't assume that just because there isn't enough drive space that certain shots must be left out. It is the assistant's responsibility to find the drives, connect them, and capture everything as required.
- The bins can be named based on tape name, and the editor will determine where to put the shots based on content later.
- Watch the audio and video levels! Distorted audio and blown-out video can come back to haunt the project at the finishing stage.
- Learn how to read a vectorscope and waveform monitor or risk being bypassed by those who can!

Drives

Understanding drives is crucial to making sure the captured video can be played back. There is a setting called Drive Filtering under the General Setting. If this is on, only drives capable of playing back the selected resolution are available. Unfortunately, when people use non-Avid drives, they must disable this setting all the time. This is because when the system does not recognize the firmware loaded onto the media drives, it assumes the drive is incapable of higher-resolution playback. This is not always true, but it is safer than assuming every drive can play back every resolution. Not paying attention at this stage can mean having to copy huge amounts of material to the proper drives. Generally, however, if there is a mismatch between resolution and drive capabilities, there will be an error about either the video or audio overrunning its buffers.

- If you can, split the audio and video to separate drives, not separate partitions of the same drive.
- The drives should be named so there is no confusion between what is a new drive and what is just another partition on the same drive.
- Never overfill the drives. Leave a minimum of 10–15 percent free space on any partition.
- Have a thorough understanding of SCSI principles (terminators, IDs, etc.).

- Be sure you can recognize the difference between drive types and speeds so that you do not slow down the performance of all the drives on the SCSI chain by adding a slow drive in the wrong place. Have the right SCSI terminators, cables, and adapters.
- All drives should have unique names and physical labels so they can be moved to any system and still be identified.

Understanding drives also means knowing how to resuscitate an ailing one and knowing when to call it quits and get another. The number of drives returned to Avid with nothing wrong is astounding. If you can get a drive back to full health in an hour or two, how does that impact the production schedule compared to waiting for the morning rush delivery?

- Know the replacement policies of your non-Avid drives.
- Learn how to use all the Avid drive utilities and make sure you have the latest versions.
- Don't always load the newest firmware until you know all the ramifications for your configuration. Read the Release Notes if in doubt.
- Know how to mount, repartition, and, as a last resort, perform a destructive read/write test.
- Dealing with very large media files still requires different rules for maintenance. Don't assume you can run any drive utility on a media drive.

Media Management

All the media on all the drives is under your jurisdiction so knowing what to keep and what to discard is both incredibly important and commonplace.

- Know how to lock and unlock media files and delete precomputes.
- Know how long it takes to copy media from one drive to another.
- Delete media through the Media Tool, Media Manager, or drag entire projects to the trash in their media folders after using MediaMover. Have a regular plan to delete precomputes.
- Understand the network and how it helps you to move media efficiently.
- Learn MediaMover and the Unity Administration tools if your facility has them.
- Become the network expert if you can and research the possibilities.
- Figure out how to improve network throughput.
- Reduce the number of drives that must be moved.

Basic Maintenance

If you aspire to become an editor, there are people who will hold your technical expertise against you. They think you cannot be technically proficient and a true artist. You may have to work a little harder to prove them wrong. It takes only a couple of success stories where you save the system and the project before employers see the value in an editor who reduces his or her own downtime.

- Take a Macintosh or Windows support class or a basic troubleshooting class.
- Consider becoming an Avid Certified Support Representative (ACSR).
- Always have a floppy disk or some other removable media that you can boot from and run disk-recovery programs.
- Be ready to do a clean reinstallation of the system software.
- Between projects do everything you can to make all the systems as similar as possible.

Backing Up

Consolidating is the most important feature for backing up. Some people have enough time and tape to back up everything in a project, but more likely, you will be forced to decide what to keep and what to discard.

- Learn all the variations for consolidating the final sequence.
- Know how to back up only the media needed to recreate the sequence.
- Create a database that can both retrieve the project and the individual bins.
- Print out the bin for each tape and include that with the tape. You will have a paper archive when all else fails.

Output

All forms of output are important and critical to the next step in the project. The digital cut may be the master or the approval copy. If it is the master, then the levels must be perfect, and if it is just a VHS or one-off DVD for approval, then it must contain all the video tracks of graphics and all the audio, mixed or direct.

- Do you have a timecode generator that can burn in the sequence timecode to the digital cut?
- Spot-check the EDL (edit decision list) for accuracy.

- A cut list should be scrutinized at every cut because the stakes are so much higher and the chance for adjustment at the next stage more remote.
- Make as many EDL versions and as many disks as you have time for.
- Expect that, when the online assembly finally comes, someone may ask for something more. If the request is for video-only EDLs, make a few with audio, too, just in case.
- Make several printouts to cover yourself, save time, and help the online editors if there is a mistake at their session. Everybody makes mistakes and, if you have an original copy of the EDL on paper, you can isolate the mistake more quickly as something done wrong after your handoff.
- Talk to professional sound studios. What are the most common mistakes you are likely to make when handing off media and projects to a ProTools session? What is correct for one production company may be incorrect for another—maybe because the other facility handles your files wrong!
- Document all the steps you take to prepare the files and label everything clearly.
- It really is part of your job to prevent other people from making mistakes!

Recapturing

- Does every shot need to be recaptured? It is possible to make an EDL and a cut list with media offline. You can even relink to master clips that are offline just for the correct version of metadata.
- If the changes requested pertain only to the open and close of a sequence, then you might be able to get away with capturing only those sections. You can leave the rest of the media offline.
- If the need for recapturing is because the levels were set wrong, then the saved settings for that tape must be deleted before recapturing.
- If some of the footage needs to be captured in black and white, make sure to set the Capture Tool (Compression Tool on older systems) to monochrome. Even more important, change it back when you're finished! The third monitor shows you a color image all the time because it is monitoring what the signal looks like before it is processed. How many times have people walked into an Avid suite during capturing at low resolution, looked at the client monitor,

and said, "Hey, that doesn't look so bad!"? The monitor during capturing does not reflect whether you have left monochrome on or are capturing at the wrong resolution.

Blacking Tape

It is desirable to know absolutely everything about connecting and operating video decks. This information is beyond the scope of this book, but it would include knowledge about:

- Reference signals.
- Signal termination.
- Loop through.
- Deck-to-deck editing.
- Deck front panel input choices.
- Blacking tape (sometimes called "black and coding" in American English or "black and bursting" in the Queen's English).

After spending an afternoon with a well-known, much-decorated documentary editor, I asked about his deck connections. "What's a BNC?" he said. Although he had a great attitude toward the new technology, there was quite a learning curve involved in making him self-sufficient.

If you are not supported by staff engineers, I highly recommend that you learn how to clean the video heads. Oxide flakes off videotape and sticks in the tiny gap that video heads use to read the video signal. If you have a regular staff of technicians, then don't touch the video heads, but make sure they are serviced on a regular basis. If no one is regularly cleaning the heads on the video deck, you should take on the responsibility as routine. Cleaning the heads with a proper head-cleaning kit once a week during heavy usage is not a bad idea, and if you are using a consumer deck, of course, there are head-cleaning cassettes. Consumer-quality tape loses oxide faster than professional tape.

- There is a significant difference between digital and analog cleaning procedures, so make sure you know the deck you are working on.
- Professional digital decks do some self-cleaning.
- Do not clean with alcohol. Alcohol leaves a gummy film when it dries.
- Use a freon substitute and a wipe that is recommended by the manufacturer.
- Some rental facilities cover the edges of the VTR top cover with a seal so they know if it was opened and tampered with. Check with the rental company before breaking the seal.

For connecting Betacam SP decks, or any decks that have both composite and component input, I recommend looping the black-burst signal out of the reference input to the composite input. This serves several purposes, but the most important is that you will never accidentally use the composite input for your digital cut. It is always a black signal. The other benefit is that you can change the position of the switch on the front panel of the deck that changes the deck input from component to composite. Then you can begin to black tape without disconnecting or reconnecting any cables. Always monitor the output of any deck that is recording or you may end up recording black when you don't mean to!

If you buy tapes by the case, a good practice is to black them all whenever there is any downtime. By turning down the audio inputs and switching over to composite video input, which has been connected to a source of black, you can black tapes at a moment's notice throughout the day. Be sure to turn the audio inputs back up (or pop them back into the preset position) and throw the front panel switch back to YRB (component) before you start a digital cut!

When blacking tapes, set the four switches under the front panel of the Betacam deck to:

- DF (drop frame) or NDF (non–drop frame) if in NTSC.
- Internal.
- Record run.
- Preset.

This ensures that the timecode is generated from the internal timecode generator. It is not looking for an external signal. The timecode will only increment (run) when the tape is recording, and the starting point is wherever you preset it to start, no matter what other timecode is on the tape.

The ability to preset a starting number for timecode is achieved by manipulating the buttons on the outside of the front panel. Each deck is slightly different, but on Sony decks other than the UVW and DSR series:

- Press the Hold button.
- Use either the + and − buttons or the search knob to choose hours, minutes, seconds, and frames.
- Press the Set button.

Although the number in the LED does not show it yet, when you begin recording, the timecode starts incrementing from the desired preset number. The UVW and DSR series use a simple menu system to set timecode.

Other switches to check on a Betacam SP deck are on top of the front panel (as opposed to on its face or hidden by it), where the record inhibit switch is also located. The 2/4 field switch can sometimes be the cause of field-based problems when capturing graphics from tape. You should always capture everything with this switch in the four-field position (NTSC).

If you are trying to lay four discreet channels of audio to the deck, you'll need to switch the front panel audio switch that says "1/2 | 3/4" to 3/4. If it's on 1/2, then the audio sent to 1 and 2 is automatically laid to 3 and 4 if those input levels are up. Many people use channels 3 and 4 on the Betacam SP because of the superior signal-to-noise ratio. They may want four discreet channels from the Avid system to the deck so the stereo music can go to channels 3 and 4 while the narration and sound bites go to channels 1 and 2.

A final note of caution: Tracks 3 and 4 are recorded helically, alongside the video. If you make a video insert edit on a Betacam SP master, you will wipe out tracks 3 and 4 for the length of the edit!

Even as Betacam SP decks are replaced by DV or other digital decks, these principles will always stay the same. Be familiar with how to set timecode on any deck you work with.

There may be more specific requirements for assistants at your facility, but these should cover the basic skills to make you useful from day one. All markets are slightly different with different terminology and different kinds of clientele. Every chance you get to watch the editors work should be snatched up, not just to see how they use the interface, but also to see what they are doing with it. Observe the way creative ideas are thrown around, accepted, rejected, modified, and experimented with. This is always more important than the equipment being used.

You have an advantage over assistants in the past because, if you can get permission, you can take the opportunity to cut your own versions of the scenes, commercials, or segments when the suite would otherwise be empty. You can't mess up the film or add wear to the tapes just dissecting what the editor did and trying a few variations of your own. If you are lucky, you can get the editor to check them over and offer suggestions. A recommendation by the editor may move your career along faster than anything else.

With the more affordable Avid systems available to you to learn with, you should always be practicing when you can. Get time on a system, even if it is at home, to improve your skills as a storyteller or your command of effects. In the past you would have had to come into a production company in the middle of the night to get your hands on the "big iron." Now you can be creative and work on your editing chops on a laptop. Take advantage of this wonderful gift of evolving technology to improve as an artist. Positions change and so does technology. If you stay flexible, always willing to learn something new, you will always stay valuable to your employer. This book outlined some of the more important techniques to keep in mind with the Avid editing systems, but by no means all of them. To keep up with the speed of change takes more effort and more research than you might have expected. Stay focused on the most important aspect of the technology—the storytelling—and you will never be out of date.

PREPARING FOR LINEAR ONLINE

"You don't really need modernity in order to exist totally and fully. You need a mixture of modernity and tradition."

—Theodore Bikel

Most of the preparation for a linear, tape-based online happens at the end of a nonlinear project when you make the edit decision list (EDL). But there are things that must be done properly throughout the job in order for the online to go smoothly. The reality is that with all the twisted financing that goes on in this crazy industry, there are still reasons to buy a bunch of offline models and create an EDL when finished. However, the days of saving money by going to a linear suite are swiftly ending. Throughout this process keep one thing in mind: You are stripping valuable metadata from your sequence whenever you make an EDL. With the growing acceptance of AAF (Advanced Authoring Format) for transporting all of your sequence's valuable parameters, you may want to think hard before going this traditional, limited route to finishing your project.

This chapter discusses the most basic requirements to get a good EDL under the most common scenarios. Then it goes a bit deeper into the possibilities for increasing the speed of an online auto-assembly, the Holy Grail of all expensive online sessions, and almost as tough to find! But the good news is that many facilities want the list in a simple, bulletproof form so that they can do their own, more complicated variations.

How do you transfer a 24-layer effects sequence with nesting, color effects, audio equalization, and rubberbanding to a format that supports one layer of video, four tracks of audio, and hasn't changed much in almost 20 years? Answer: Not very well. Dedicated hardware is the rule in traditional online suites with a different piece of equipment and, many times, a different manufacturer for each function, like text, special effects, and deck control. The separate pieces of equipment were purchased years apart and there may be components that are 10 years old that still work fine for what they were designed to do. Compare the processor of a CMX 3600 with a high-end workstation Macintosh or PC. The workstations have consistently supported Moore's law and doubled the amount of processes on a single chip every 18 months. Yes, if you design the hardware to do just one thing, then you can maximize it so that it doesn't need as much RAM or processor speed. But just try to add new features! This is why there are so many third-party EDL management software choices. If the dedicated linear online systems cannot add new software features, then new third-party software does it and simplifies the EDL to a format the dedicated systems can handle.

Even with third-party EDL programs, there is still a massive mismatch in capabilities between what can be done by clever folks with the Avid and what can be represented by this limited format for a linear online. The only way to convert from one format to the other, preparing for that precious linear online time, is to dumb down everything. You need to understand the limitations of what can be conveyed in an EDL. Bring the Avid EDL to a point where the show can be built back up again, piece by piece, in the order necessary in the linear, tape-based world.

This is more of a challenge than it would appear at first, not because it is that difficult, but because of the limited experience of the Avid editor in the world of linear online. Unless you have actually worked as an editor both at an Avid system and at that particular linear online suite, you are not really in a good position to take advantage of all of that tape suite's advantages. We can discuss capabilities in generic terms, but that doesn't help you for the specifics of that edit suite. You need to know that for that suite, you needed to reserve the character generator a day in advance and you can forget using the DVE (digital video effects generators) unless you book a night session.

This is where you must rely on two very important low-tech tools: clear communication and the ability to collaborate with experts. As you devolve the Avid list to something simple for the linear assembly, you must be very clear about exactly what must be done to every part of your sequence. What effect do you really want? How are you going to deal with all those channels of audio? This chapter will describe different ways to leave a paper trail that someone else can follow. Count on the expertise of the operations or scheduling people at the particular postproduction house that has been chosen. Speak to the online editors if they are available. Depending on time available and the complexity of the program, you may even want to (gasp) plan a pre-postproduction meeting to discuss approaches with the editor! Imagine, warning him or her of what to prepare for! Let's face it, there is no way to predict everything that can happen after you leave a project. It is really not fair for the offline editor to be blamed for changes that occur after the cut is locked. Sometimes changes ripple backward to affect a decision you made in good faith, with the information available weeks or months ago. Unfortunately, this happens all the time.

Coming from a postproduction facility background, but having freelanced for a few years as well, I strongly advocate building a good relationship with a handful of production companies. Bring any work you can to these select production companies, large projects and small, and spend some time learning the facility's capabilities and quirks. If you create a sense of loyalty, they will look out for you. This happens in little ways, like working a little harder to fit you in for emergencies. They may come to know your work and style and be able to anticipate and fix problems for you before you even know they are there. If they can get you in and out on time and on budget, they can fit in more paying sessions, need fewer "make goods" for questionable mistakes, and lower their already unnecessarily high stress level.

If you can afford to sit in on a few online sessions, it will pay off for your customers in the long run. If they have a relationship with you, then it is in their best interest for you to continue to make informed decisions about effects that can be recreated easily, or faster ways to assemble the show in the final stage. Will it be faster to dub shots onto a selects reel? Is it better to have more tape decks available for a shorter time or fewer decks for a longer time? And really, what can you do at a lower rate per hour to make things go smoother? When you learn these things, you become more valuable to your employer and have more job security. You are more likely to stay on the A-list the next time your favorite client, director, or producer has a big job.

So it is in everyone's best interest that you understand the implications of every step of your offline on the final online. If you capture without timecode or without paying attention to tape names, you are directly hindering the next step. Don't blame "those Avid EDLs" when you have no regard (and, really, no respect) for the next link in the chain. It is your job to continue to inform the director or producer that choices they are forcing you to make will cost them money later. If they take just a little more time now, at the lower rates, things will go smoother at the crunch time.

What is the biggest mistake? Ignoring the importance of timecode and tape names. Everything must have a timecode reference—all sound effects and music, all graphics and animations, and all picture and sync sound. Let me say this again: *If you are going to a linear online, every source must have timecode.* It is tempting to import just that one cut from the CD music library or pop in just those three shots from the VHS dub. But you must be sure to go back and match by eye or ear with a timecoded original source before you make the EDL. This is a clear example of GIGO—garbage in (sources without timecode) and garbage out (a list that cannot really be assembled).

Preparing Sound

There is one exception to the timecode rule: when the audio captured into Avid will actually be used as a source in the online (or the final mix for a film). It is becoming more common for the first edit to lay down a digital cut of the audio, low-resolution video, and then cut with original video sources the rest of the day. This assumes no changes in the online that will affect sync or length; of course, this is something you may have to mention repeatedly! In this case, you are working from the timecode of the digital cut, which should match the timecode of the master sequence. If the show starts at 1:00:00:00, then the sound source should, too. For transferring just the audio from the offline session, a timecode-controlled DAT or another digital tape format like the eight-channel DA-98 works quite well. You could go out of Avid using the digital audio output format to a Digital Betacam or another video format that supports the digital material you will be outputting. In a pinch, audio tracks 3 and 4 of a Betacam SP can be used since these audio tracks have a higher dynamic range and lower signal-to-noise ratio than tracks 1 and 2. This is good for effects and sound under, but should not be the main method for the most important sound in the project. Just make sure the Betacam SP decks in the online suite are models that can play tracks 3 and 4. Generally, this means a Sony editing deck with the letters BVW in front of the model number.

Did I mention you can't make changes?

Using the Offline Cut in Online

If you are using digital videotape to bring the Avid digital cut to the online suite for your audio, then you should also include the video as part of that first edit. Many online editors use a clever trick where they lay down both video and audio from Avid to the master tape in the online suite. They put a big circle wipe with color bars in the center of the video. That way they can cut over the top of the old video all day and see if there is any discrepancy between the EDL and the digital cut. Any time they see a flash of color bars they know something is a frame or a field off and they can make adjustments. It sure is better than squinting all day to tell the difference between one frame of 10:1 and the original!

The other way that the captured audio without timecode from Avid can be used is when you make an OMFI export of the audio and the sequence for use on a digital audio workstation. But if the original sound was captured with a little distortion or mono instead of stereo, then the sound mixers have to recapture it on their workstation and match it back by ear. This is not too much trouble if it is just a few cuts, but if the music comes from a five-minute cut and is used at every transition, then you are costing your client money. How do you think the sound guys will justify their higher bill?

When you are dubbing everything to timecoded sources, it is up to you (or your assistants, dubbers, etc., but really you) to make sure the quality is not degraded. Dub to higher formats—digital formats—and monitor the levels very closely. There is nothing worse than being forced to recapture a distorted media file and finding that the dubbed source is also distorted! Some facilities make sure that everyone who dubs a source puts his or her initials, date, and machine used on every tape. Being able to track a problem back to the source does a lot to ensure quality control.

Dubbing with Timecode

There are some basic requirements for good timecode when dubbing sources to be used in an online session. The first requirement is: Never dub timecode. That seems like an outrageously stupid assertion, but one that has confused people for many years. You should not dub timecode—you should always regenerate it. Timecode is a square-wave signal, and straight dubbing tends to add noise to the signal. Eventually (sooner if the timecode wasn't great to start with), the signal becomes slightly rounded and suddenly, where that sharp edge in the signal delineated a specific number, it is now just a little too rounded to read accurately. You may have been dubbing timecode for years and sending your problems downstream or, luckily and more probably, your record deck has been left in the position for Regen. This is a switch under the front Control Panel of the deck that allows you to regenerate—resquare—the signal as it passes through the deck's electronics and lays it to tape.

It may be important to keep the timecode from the original source for accounting reasons, so the stock footage can accurately be paid for or because of the way the selects reel will be made. When making a selects reel, you can be dubbing many shots from separate reels onto one reel for the convenience of capturing or because you will lose access to the originals during postproduction. The shots on the selects reel may need to be traced back one more step to the original tapes before the online. Now you should set the internal/external timecode switch on the tape deck to ext for external source. Make sure that the source deck and the record deck have timecode cables connected between them so that you can jam-sync the timecode to the new tape. Give yourself plenty of preroll because there is a break in the timecode at every shot. You do not want the Avid or the edit controller to rewind over a break in the timecode when cueing up for a shot.

All edit controllers, whether they are part of the Avid software or part of an online suite, always expect a higher number to come later on the tape. The timecode is expected to increment continuously! If you put hour 10 before hour 5 on the tape, anyone who uses this tape will be confused. Edit controllers will search and fail to find preroll and edit points. Your name will be used in vain and no one will ever return your phone calls again.

If the beginning shots of the tape are drop-frame, then everything that is dubbed to that tape should be drop-frame, too. The Avid system associates a tape name with whether an entire tape is drop or nondrop. If you mix the types of timecode, you have to give the tape two names—this is very confusing and not recommended!

If you can lay down new timecode instead of taking it from the original source, you can keep the timecode nice and neat on the selects reel. If the ext/int switch is back in the int (internal) position, the record deck always generates its own timecode. Another benefit of leaving your deck in Regen is that any assemble edit (an edit that changes the timecode on the tape) picks up where the last timecode on the tape left off. During the few seconds of preroll prior to the assemble edit, the deck reads the timecode already on the tape. At the edit point, the deck neatly regenerates timecode without missing a frame. Either you should be purposefully presetting the timecode on the deck, jam-synching the code from the source tape, or regenerating it from the timecode that is already on the new tape.

Timecode is just another tool and can be used very effectively once you know how to preset it and actively use it. If you are making a selects reel, choose a timecode that is not used by the other field tapes. If the last field tape is tape 15 and uses timecode hour 15, then start your selects reel at hour 16. No matter how a tape is named in the EDL, you can trace it back to the right tape based on the hour of the timecode.

Tape Names in EDLs

The tape name is incredibly important to the Avid system and, if you get almost nothing else right, at least this should be perfect. Every tape needs a unique name. A tape must be named while capturing or logging in the correct project. It is good practice to make the tape name match the timecode hour and never repeat a tape name. As mentioned before, but bears repeating, tape names should be short—five to seven characters maximum to allow a space for a "B" to be attached to the end when making dupe reels for CMX and Grass Valley Group (GVG) systems. This chapter discusses dupe reels later—you can't count on not needing them!

Assuming that all timecodes in Avid refer back to real sources and all tape names are unique, let's look at the basics for preparing a list. Be sure that you have installed EDL Manager from the Avid installation CD-ROM. This piece of software creates EDLs from Avid sequences and is meant to be able to open bins and locate the sequences inside the bins to create EDLs.

EDL Manager

EDL Manager is Avid's EDL generation and list-importing software and is separate from the Avid software for several reasons. The first reason is so that it can be updated and improved upon with a development schedule separate from the editing systems. EDL Manager can then skip the long and complicated beta-testing programs in favor of quality assurance just for this small piece of software. Being separate means that the offline editor can carry it to the online suite or the online editor can have a version running in the suite already. The offline editor can be spared some of the responsibility for making a good list if he or she brings a sequence bin to the online session with the (ideally) final version of the sequence.

The EDL Manager launches automatically when you choose "Output > EDL." If it does not launch then perhaps either you have not installed it from the original installation DVD-ROM or someone has cleverly hidden it somewhere on your system and now it cannot be found. Instead of troubleshooting, it is quicker to go back to the original DVD-ROM and reinstall the application in the proper location.

If both applications are running, you can use the simple arrow icon in EDL Manager. The sequence, which is in the Record window of the Avid editing software, will load into the EDL Manager with a single click. You can also use the arrow to go the other way and turn an EDL into a sequence, as we discussed in Chapter 9.

Getting Ready

There are so many variables in the linear online suite that you can never be too careful about preparing for any eventuality. There are many different kinds of edit controllers on the market today and, to some extent, they are not compatible. The main ones in use are CMX, Grass Valley Group (GVG), and Sony. There are also many other less-prevalent models like Ampex ACE and Abekas; however, within your market there may be an exception to this, and you may find that the most popular edit controller is something else. In Chicago, for instance, you will find many Axial controllers.

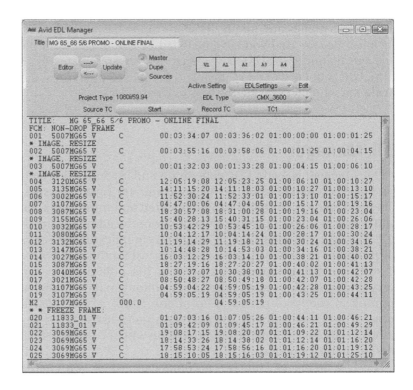

The first rule of EDLs is this: Find out which edit controller format the linear online facility needs. Notice I said what they need, not what they have. This is for a very important reason: They may have a CMX Omni, not a CMX 3600, but the lists are mostly compatible, so the facility may ask for a CMX 3600 list instead. Or they may have a bizarre ten-year-old Brand X model that went out of business seven years ago, but the darn machine just refuses to die. They may ask for a CMX 3600 as well because out of necessity they may have, or may have written, their own software to translate from CMX 3600. Most people consider the CMX 3600 the Latin root of all list formats.

Formatting Floppies and RT-11

The biggest challenge you'll run into with EDLs is that edit controllers are primarily old systems that take *dual-density floppy disks(!)* as their input. Needless to say, you may have a problem providing your list on an EDL considering that neither modern Windows or Macintosh systems have shipped with floppy disks in some time. You may even have trouble finding blank disks at your local office supply store. For this reason many postproduction houses now accept lists on CD-ROM or USB memory sticks. They'll then transfer your list to a floppy. But if you want to do it yourself, you can purchase USB floppy disk drives for either platform—at least as of the publication date of this book! Floppies are a dead technology in the general computing world.

If you do want to use a floppy to submit your EDL, you must format the floppy disks for CMX and GVG edit controllers in an unusual way. CMX originally used a disk format that was useful with the DEC computers they used internally when they were developed. Because GVG developers came from CMX, GVG also uses a version of that original CMX format. The format is called RT-11 and is problematic with both Macintosh and Windows because these operating systems are not designed to recognize it. Special commands must be written to make the computer ignore the fact that it can't read the RT-11 floppy disk you just popped in. The edit list software must take over for both reading and formatting these floppies. This is why if you just pop a CMX disk into the Macintosh, you get a scary message saying this disk is unrecognizable and asking if you want to format it. No! Eject, now! The Mac defaults to this evil question whenever it cannot recognize the formatting on a disk. You can work with RT-11 disks only when you are in Avid, EDL Manager, or another piece of specialized software designed specifically for EDLs.

RT-11 was not written to make your life a living hell, but it sure doesn't help. Add to this the restriction that the disk used must be a double-density, double-sided disk. I want to say that again: double-density, double-sided. The only exception for CMX and GVG is the HDGVG option in EDL Manager, which will correctly format a high-density (HD) disk. Look at all the floppies strewn around your computer (if you still use them). I'll bet none of them used for standard, everyday file saving is double density, double sided. They probably all say HD, but the disks you want should say DD (double density) on them. They are not interchangeable!

This is an example of dedicated hardware being passed by very quickly by technological standards. Double-density disks hold only 720 KB of material. This could be a small novel, but it is not very many EDLs. The real world uses HD floppies because they hold 1.4 MB, almost twice the capacity of the DD disks. Even the HD disks are pretty pathetic in a world where gigabytes are becoming the standard measurement for storage capacity. If you are making the edit list, it is up to you to scrutinize every disk that is handed to you.

I strongly recommend that the DD disk should be unformatted or already formatted for RT-11 from the edit controller that will use it. The formatting of the disk for the EDL has caused many people the vast majority of their EDL woes. On older Macintosh systems you will need to "unformat" a floppy. First, quit all the programs you are running. Pull down the Erase Disk command under the Macintosh Special menu, as you would normally erase all the material on the floppy. Let the process run for a few seconds, but before it finishes you must force the computer to crash. The only way you may be able to interrupt the erase is to press the Restart button on the outside of the computer case, or on a Macintosh, press the Ctrl-Command-Power keys all at the same time. This works to totally mess up the format on the floppy, but it sure is ugly. For all of the Windows versions of EDL Manager and later versions on Macintosh, this method is not necessary since the system will competently erase any original formatting and replace it with RT-11.

So now you have an unformatted double-density, double-sided floppy. If you pop it back into the Macintosh, the system tells you the floppy is now unrecognizable—this is one of the few times you want to see this message! Windows will give you a similar message. Now the floppy can be reformatted while you are running the EDL Manager or other EDL software.

Is it any wonder that busy postproduction facilities would rather FedEx you a few of their own disks? Overnight mail with a proper set of disks is a good idea from their point of view, so give them the chance by letting them know you are coming far enough in advance. Sometimes there are differences that have developed over time between the disk drives at their facility and your brand-spanking new Apple floppy drive. Old drives get cranky and occasionally may not accept a floppy that has been formatted by another drive. The busy facility also has seen more than their share of HD disks with EDLs and offline editors who blame Avid! If all else fails, to get your floppy to be accepted by their online edit controller, copy it to another computer that can read RT-11. This other computer potentially may have EDL Manager running. Turn the EDL into a text file using a word processing program, and copy it to a properly formatted disk.

Sony and DOS

One way to avoid all of this is to work with a Sony list. Sony saves to a DOS-formatted disk that the Macintosh can easily handle. Early versions of the Macintosh needed one of two different extensions, either PC Exchange or DOS Mounter. These extensions need to be enabled in the Extensions folder to read or format a floppy for Windows. Choose "Other" during the formatting dialog on the Macintosh and choose the proper format for the size floppy you have. Early Sony edit controllers may read only DD disks, but the later ones read HD. If in doubt, get the DD and format 800 K.

Real-Life Variations

For one final twist, there may be weird combinations of formats for disks and formats for edit controllers. The later CMX controllers, like the OMNI, can accept DOS-formatted disks. The production company may have its own list optimizing software that runs on a DOS or Windows system, so they may want the list, no matter what it is, on a DOS floppy. There used to be edit controllers that worked from 8-inch floppies; most people have never even seen these before. Forget about compatibility. These unlucky people have devised a way to send information serially, through a cable from another computer that can handle the 3.5-inch floppies to their online system that can handle only 8-inch floppies. EDL Manager allows this, too, but it takes a 1-2-3-Go! procedure that you probably don't want anyone to see.

Be Prepared

There appear to be many ways to get this whole EDL process wrong because there are so many ways to get it right. With all this potential for a 9 AM thrash at the online facility, there are ways to cover yourself. First and foremost, you should ask the production facility what they want. If you are using the same production houses repeatedly, then you know their drill and can have procedures in place for each edit suite of each production company you use. There is a very useful feature on newer EDL Manager versions that allows you to make a template for your favorite edit suites.

EDL Templates

Regardless of whether you are an offline boutique that sends the lists to another facility or you are the facility and must make different EDL versions for each of your edit suites, you could use a little help. To make sure the lists are always correct and to make sure that you can turn this chore over to even the least-experienced assistant, you can make EDL templates. These templates can actually be made by the postproduction facility and emailed to you as an attachment. You would then place the template file in the EDL Manager Templates folder inside the Supporting Files folder of the EDL Manager folder. These templates are simple text files, so they can easily be copied to a floppy disk and loaded onto any system with the proper version of EDL Manager. As a facility, instead of faxing a complicated series of instructions, you can email or overnight mail the small file on a floppy disk, which will guarantee that the offline editor gets it right.

You will also find that you may need to make multiple versions of every EDL for different reasons. You may need an audio-only EDL and an EDL for each of the video tracks used in the offline sequence. If you create a template for each of these often-used versions, you can make them all much faster. Just choose the Template pulldown menu for Video Suite 1 Audio Only or GVG Suite Video 2 and quickly save the new version. With a preformatted floppy and a template from the postproduction facility, how can you go wrong? But don't bet everything on one floppy—use two, they're cheap! You may also want to save the EDL in several other formats. Always make exactly what the facility asks for and then make a CMX 3600 list. Pick your market's favorite format and make that your backup format. It is up to you whether you want to put it on DOS or RT-11. (If you choose both, it could make you look pretty slick.) Then you should burn the sequence bin to a CD-ROM and, if the project was small enough, include all the bins. You may find yourself actually making the list again from the sequence at the online facility's Avid or from your own copy of EDL Manager (which you brought with you). After this is done, trust nothing! Make two printouts and bring a copy of the project on VHS or some other format that can be played back in the online suite without a charge for another deck. An MPEG file on a CD-ROM could work, especially if you have your own laptop to play it back. You or your producer should control the playback of this copy in the online suite so you can follow along with the online assembly.

What Is an EDL?

Now that you are prepared to make the EDL, let's take a look at what an EDL really is and why it is so hard to make one from a nonlinear editing system. The only information the edit controller wants from Avid is the tape name and timecode (sound familiar?). The edit controller cues up the source tape and inserts the shot onto a specific timecode on the master tape. So a basic event—a cut, for instance—needs four timecode numbers: an In point and an Out point on the source tape, and corresponding In and Outs on the master. It looks like this:

```
017 100V C 02:00:24:19 02:00:25:14 01:00:22:02 01:00:22:27
```

When you need to make a dissolve with analog tape, most of the time you need a second source. If you must dub a shot onto a second tape, typically it is called a B-roll. Thus, an A/B-roll system is one that allows dissolves. A dissolve needs the In points and Out points from the B-roll in the event, and looks like this:

```
012 100V C 02:18:31:29 02:18:31:29 01:00:12:19 01:00:12:19
012 100B V B 030 02:23:26:15 02:23:28:26 01:00:12:19 01:00:15:00
*BLEND_DISSOLVE
```

Any kind of fade up from black or from a graphic that has no timecode is an event that starts with black (BLK) or an auxiliary source (AX or AUX). It looks like this:

```
002 AX V C 00:00:03:27 00:00:03:27 01:00:01:28 01:00:01:28
002 100V D 030 01:43:44:28 01:43:50:29 01:00:01:28 01:00:07:29
*BLEND_DISSOLVE
```

All edits that come from a nontimecoded source, like CD audio or a PICT file, come in as AUX. If you have 100 PICT files, this could be a problem.

Translating Effects

Any time you create an effect like a wipe or a three-dimensional (3D) warp, it must somehow be translated to this very limited, older format that is concerned with tapes and timecodes. There are standard SMPTE (Society of Motion Picture and Television Engineers) wipe codes, but some of the most popular effects switchers do not use them. EDL Manager has a list of the most popular switchers represented, and you can use these simple templates to represent your wipe effects. You can even take these templates for the switchers and modify them to your own switcher if you are feeling geeky. The wipe codes can usually be found in a manufacturer's technical manual, and these templates are just text files.

There is no way to create a generic 3D effect that any group of DVEs made by different manufacturers can read. Many people don't have the capability to send DVE information from their edit controller to the DVE. These two machines easily could be from different manufacturers and have no communication between them except "Go Now." Different DVEs use different scales for computing 3D space, their page turns look incredibly different (if they can even do a page turn), and really, forget it, there is no standard for effects. There won't be such a standard for a long time to come. You have the timecodes for the two sources for the effect and you have the VHS. If you are really persnickety, you may even have written down some of the effect information as you made the effect. Adding a comment to the list like "second keyframe is exactly 20 frames later" doesn't hurt.

This information is best delivered to the EDL itself using locators or the Add Comment feature in Media Composer and Symphony. You can use the text entered into locators as comments in the EDL. This is an option in EDL Manager's Options window under Show. To use the Add Comment feature you highlight a single shot in the sequence with the segment arrow, pull down the menu above the Record window, and use the Add Comment option. The comment shows up in the EDL as a little bit of extra text when you choose "Show: Comments" in EDL Manager. You can search by locator in the locator window or by comment in the sequence using Ctrl+F/Command+F in the Avid timeline.

Even if you can represent exactly the way an effect was going to happen, you cannot foresee how the online edit suite will be wired. The effect may call for three levels of dissolves, but the switcher has only two levels it can do simultaneously. Your effect may require that the editor create multiple layers of video, which the editors should know how to do quickly and efficiently with their own configuration of equipment. If you try to be too specific about the exact procedure for creating an effect, you are denying the online editors the chance to do what they do best.

Multiple Layers of Graphics and Video

The final insult in translating from nonlinear to linear is the way most EDLs represent multiple layers of video. They really don't. This means you need to reduce all multilayered sequences to a series of single–video track EDLs. The nested tracks must be broken out of track one and put into separate EDLs for each track in the nest. This provides you with the ability to run more than two video sources simultaneously in the online edit suite. Don't count on it unless you have researched the specific suite you will be using.

The online editor loads the video tracks as separate sequences into the edit controller and combines them. You can help the online editor figure out which tracks go where by the way the separate lists are numbered. Each edit is called an event and has a number. If the first layer of events ends at event 110, then all of the events for the second layer of video can start with event 200. The third layer can start at event 300, the fourth layer at event 400, and so on. The editor knows that when two events overlap in the EDL and if one has event number 50 and the other is event 210, then 210 is part of a multilayered effect. Any multilayered sequence should be discussed carefully with the individual online editor. Make sure they can even get close to what you are doing on the Avid. You may find they say, "Sure, we can do that—in the compositing suite!" Listen for the cash register sound when they want to put you into a special compositing suite. There may be no other choice but to go this route. On the other hand, a slight redesign and some extra goofing around may save you thousands. A freelancer using Adobe After Effects or another desktop compositing program may be the right answer if you have the time to render. There is a tradeoff between waiting for a short render in the expensive compositing suite and waiting until Monday for After Effects. With the faster computer processing units (CPUs) the render times on the desktop are becoming more competitive. It may make sense for you to get a cheaper online suite and arrive with all the effects already made and output to tape so you just slug them in. Did I mention that you can't make changes in the online suite?

Simplify the EDL

If you try to make an EDL from a complex sequence with layers and nesting and you just load it all into the EDL Manager and press the purée button (or Update), what comes out will not be very useful. You may get some error message as the system warns you that what you are trying to do is not very nice. This is usually some kind of a parsing error or a message that an area of the sequence is too complex to be represented. Make a copy of your sequence and start to simplify.

First, get rid of the nesting. Figure out what are the most important sources for that effect. What timecodes do you need? Keep just that source. If you absolutely must have the timecodes from multiple sources for that effect, you should not have nested it in the first place. Yeah, right, that assumes you knew you were going to an EDL! If you are going to an EDL, do not nest. Enter the extra timecode and source information by hand as a Comment or a locator. In EDL Manager, choose which video track to use, starting with V1.

If you know you are going to an EDL, don't use imported graphics. An online suite may be able to use PICT or TIFF files through a sophisticated character generator, but if you know you're going to an online room, all graphics should be fed to you on tape. The timecode numbers of the fill, matte, or backgrounds can go cleanly into your EDL with few questions asked. Also, if you use imported graphics, it means additional work for the graphics department because they must output the electronic version for you, then they have to make a fill and matte and lay it to tape for the online. If you have subtle variations in logos or graphics, the online editors can put in the wrong graphic since now you expect them to figure out which graphic on the graphics tape corresponds to your nontimecoded source. Even if you use imported graphics to simplify your edit or because you didn't know you were going to an EDL, I strongly recommend recutting those sections with the same graphics tape that is used for the online.

Sound Levels

How did you change your sound levels when you were mixing? If you used rubberbanding, there is no problem. Those levels are not in the EDL, but the sound edits do not confuse the online editor either. If you used the older method of creating Add Edits, changing the levels and adding dissolves to smooth out the level change, then you have some work to do. These Add Edits are real edits and the EDL Manager is smart enough to ignore them, unless you have added a dissolve. The EDL Manager puts an edit in the EDL and tells the online editor to dissolve to a B-roll, a dub of the exact same audio source, just to change the audio level. This is confusing and annoying. You need to remove the dissolves to any audio transition that is just for a level change. This is actually easy because you can go into the effect mode and lasso or Shift+click them and delete them all at once. This is a good reason to use rubberbanding. You can also use a feature in EDL Manager called "Audio dissolves as cuts."

For the high-end jobs, the sound usually is sent out for audio sweetening and is laid into the online session in one big edit. For lower-cost jobs, a digital cut is laid to tape (sometimes split tracks, sometimes stereo split) along with the low-resolution video. Then the online editor cuts his or her video on top of that, covering it. For these EDLs, disable the audio tracks, creating a nice, clean, video-only EDL. To be safe, make an extra EDL with audio and video, just in case they need to find the video for a sound bite that has been covered up entirely. It will also help if there is a problem with the sound from Avid in a certain spot.

Settings

There still seem to be many choices in the EDL Manager that haven't been described yet, so don't get nervous. Most of these functions are for streamlining an auto-assemble and other more obscure situations. We will get to them at the end of the chapter, but for now let's go with the simple answers.

Sort Modes

A sort mode is a way of ordering the events in an EDL so that they can be edited out of linear order. Linear order means that you load a tape into the deck, adjust the levels from the color bars at the beginning of the tape, fast forward to the spot on the tape where the shot is, and perform the edit. Then you load the tape for the next shot in the program and do the same thing. You do this whether the source tape is 20 minutes long or two hours long. There are a lot of repeated actions and wasted time shuttling the source tapes back and forth. Unfortunately, a linear order is the easiest to follow and the safest way to assemble a show and be able to compensate for a mistake or make changes. You know immediately if a shot is missing or a sound bite is wrong because it does not make sense as you put it in place.

The time savings are significant if you can assemble a show out of linear order. It means you need to load a source tape only once, do all the color bars setup, and skip through the show, dropping shots wherever they are needed from that source reel. Then put the tape away. You might have to bring it back for effects later, but if your show is just cuts and dissolves, you have made serious savings in time and money. But imagine making a change that makes the show just a few seconds longer. If you have been inserting shots all over the master tape out of order, this is pretty serious. How can you ripple that change and push down all the edits after it in the program if there are already 100 shots laid down after the change? You can't really; if you are working with digital tape, you can clone the master and make an exact digital copy. If it is an hour-long program and you are halfway done, then kiss goodbye a good 45 minutes of setup and dubbing time. Did I mention that you can't make changes in online?

This kind of penalty for being wrong or making changes at the last minute tends to frighten even very brave online editors. For this reason, they want to be the person who decides how to sort the list. When asked, they inevitably tell you to use the A mode because that is the linear mode and they sort it themselves in their edit controller to make it a B, C, D, E, or S mode.

Let's do a quick recap of what the different modes are so you can discuss them intelligently at cocktail parties. B mode is for long-format shows where the source tapes are shorter than the master tape. The source tape fast forwards and rewinds and finds shots that go in a linear order on the master. This is easier to follow because the master moves forward while a specific source tape is being used. As soon as a new tape is loaded, the master rewinds to the first place the new tape is used. It then starts to move forward again until all the shots from that source are inserted on the master.

A C-mode list is for when the source tapes are longer than the master tape. You may have a 90-minute film transfer for a 30-second commercial. (I hesitate to call them "spots" anymore since in the United Kingdom that means pimples!) Here, the source tape always moves forward in a linear way and the master tape flails around rewinding and fast forwarding to assemble out of linear order. Potentially this is faster than B mode but harder to figure out what is going on. It is very disconcerting for clients who should just look away if they are feeling queasy.

D and E modes are like B and C modes except that all the dissolves are saved for the end. This works only if you have time base correctors that have memory for individual tape settings. More common these days is a serial digital interface (SDI) signal from a digital deck. Supposedly, the SDI signal does not change from deck to deck like an analog signal and so, theoretically, never needs adjusting. The editor must match a frame perfectly from something that he or she adjusted that morning or the day before so that you don't see where the pickup point is to start the dissolve. D and E modes get used more in PAL countries where they don't worry about analog hue adjustment, unless it's terribly wrong to start with!

S mode is useful for getting a list that will be used for a retransfer from film. If you have been cutting with a one light or a best light, inexpensive film transfer without shot-by-shot color correction, then you want to go back and retransfer the shots that were actually used. S mode is the order that the shots came on the original film reels and allows the colorist to identify quickly only the used shots for the final color correction.

Dupe Reels

After choosing the format and the sort mode, it is time to think about dupe reels. What are you going to do when you need to dissolve or wipe? If you haven't experienced the difficulty of dissolving between two shots on the same tape, then you are about to. With a traditional analog tape suite, you must physically copy the second shot onto a dupe reel so that you can roll two tapes at the same time. It seems kind of old-fashioned and quaint now, but a dupe reel is still the necessary evil in many suites across the world. The biggest change to this procedure was *preread*. Someone figured out that since the digital-format tape was just outputting digital information, short amounts could easily be held in a buffer, or temporary memory, in the record tape deck. During the preroll, the master record machine loads the digital images of the shot already on the master tape into the memory and plays it back as a source deck. With the last shot held in memory, the master tape switches from being a source and records the dissolve with the new incoming shot.

Some editors also use a kind of nondestructive preread called *auto-caching*. Auto-caching holds the end frames of a shot for dissolves and various layers of multilayered work in a cache that uses a digital disk recorder (DDR) or a digital tape. The edit controller figures out what needs to be cached for later use in a dissolve or effect. Then it automatically lays the image on the DDR or digital tape. When the tape with the other half of the dissolve is put into a deck, the controller performs the dissolve between the tape and the auto-cached piece of video. This works better than preread because prereads are destructive, meaning you are committed to that edit permanently because the last image is recorded over itself when the recorder becomes a source. You get only one chance to do a preread correctly, and then you must reedit the previous shot. Auto-caching uses a preread EDL to create the auto-caches that can be executed and reexecuted if needed.

Most of the choices that deal with how a tape is dissolved are refinements on these basic requirements for either a dupe reel or preread. Under certain conditions, you can significantly reduce the amount of tape loading and shuttling, but most of the time you choose the simplest technique. The online editor has a very specific requirement and again, as with sort modes, can modify the simplest format into something he or she can use best.

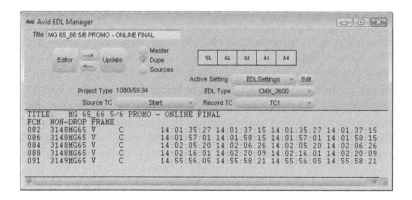

The simplest, best answer for most circumstances is the Multiple B-Rolls option. It is also the least clever, but with more sophisticated editing systems out there, editors can modify this list to speed up the assembly. The Multiple B-Rolls option assumes that there is a separate B-roll or dub of every tape in the show. This is probably not true unless you have a very small amount of material or have made simultaneous duplicates during a film transfer. It is more likely that the online editor looks at the list, sees that a dub is required, and hunts down the original tape. The timecodes on the B-roll are the same as on the original tape, and the name of the B-roll is a variation of the original tape name. Finding the shot is easy because it has a B after the tape name or the tape number is incremented by 500, depending on the type of list format. The editor can then just dub that shot when it is needed or change it over to preread in the edit controller.

Making a Special Dupe Reel

The other B-roll choices allow the editor to prepare before the online session by taking all the shots that need to be dubbed and putting them onto one reel. This keeps the tape loading down to a minimum and allows you to move faster with fewer decks configured (and charged for).

The most dangerous dupe reel choice of all is the One, New Timecodes option, although it is extremely popular in Europe. This requires that you make two lists: one for the master tape and the other for the tape where all the B shots will be dubbed. This new dupe reel has perfect new timecode because you are ignoring the original timecodes of the source tapes and starting all over again. You take each tape and dub one after the other onto this new reel, which starts with timecode that begins at 1:00:00:00 (although you can change this). This means that the timecode link between the original source tape and the dupe reel will be completely broken as soon as the dupe reel is made. The dupe reel name, which defaults to "B.REEL," shows up in the list every time a shot needs to be dubbed for an effect. You cannot look at the EDL dupe reel name and figure out where the shots came from originally. Many people choose One, New Timecodes and then do not make the dupe reel list—a big mistake. When the online editor turns around and asks, "OK, where is this B.REEL?" you really should have another list to hand over (maybe even the premade reel). Unless you make the B.REEL yourself, that reel needs to be assembled at the beginning of the session.

Because it is assembled before the editing really starts, changes to the B.REEL are also very difficult. You must make a second list for the dupe reel! If you must make a dupe reel, then One, Jam

Sync is better. If you forget to make the list for the new dupe reel, at least the original timecodes are in the list and, if the timecode hour matches the tape name, finding the original is a snap.

Another dangerous choice for the unsure is None. This gives you a list that most edit controllers cannot read, which at first seems like a rather bad choice. Not so if you are going to another nonlinear editor (NLE) or an older compositing system. These systems do not need to dub a tape in order to dissolve because, once captured, individual shots are divorced from the physical restrictions of tape. Loading the list into the NLE system forces it to see tape "001" and another tape named "001B." Is this the same tape? There is no way to know for sure, so the NLE system makes the safe but incorrect assumption that 001 and 001B are two distinct tapes. When you are capturing this material, the system asks for tape 001 and then proceeds through this tape to the end and grabs all the shots that are needed for the sequence. Then the NLE system asks for tape 001B, rewinds, and does the same thing again with the same tape! This is not efficient and, at those room rates, you need to be efficient. The None choice is the best for this situation since it shows just one tape, 001, and the NLE system requests that tape only once.

Printing the List

When you print out the list, you should make two versions. Once you are ready to print the EDL, think about who will be using the printout. The people supervising the online edit certainly need it in front of them. They may not be the most list-savvy people in the production, so there are ways to make the list more readable. Include comments in the list when you are about to print. Use comments like "clip name" and the source table, which is a list of all the sources needed for the online assembly. Then print a stripped-down version for the online editor. Again, ask if there is any special information the editor wants, like audio patch information or repair notes. I know some offline editors are reluctant to have repair notes show up in their list because it looks like they did something wrong. Repair notes are important because it may just be a case of the effect being too complex and not being able to represent it correctly. If you have any repair notes, then the online editor should see them.

The Source Table

The important function of the source table is when occasionally the EDL Manager finds a tape with a name that is too long and must shorten it. Sometimes there may be a duplicate tape name from another project. There may also be two separate versions of the same project that have been combined. The EDL Manager has to change that tape name in the list. The source table tells you that since you had a tape 001 from project X and a tape 001 from project Y, that with a Sony format, EDL Manager changed the second tape name to "999." A CMX list changes a second tape 001 to "253." According to the system, these are two different tapes! This is why the original logging is so important. If you don't have that information, then you could get very lost. Never attach a source table to the electronic version EDL on the disk because it is really just a very long comment added to the last event. Many systems have a limit as to how many comments they can read per event, and this will cause an error when reading the list if you have a lot of tape sources. Make the electronic version without the source table, but always print it out separately to take with you.

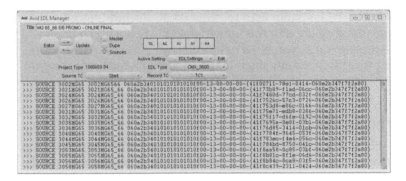

You can truncate the names of your tapes in a very specific way. This is important if your tapes have been given names that do not match the restrictions imposed by the EDL format. Most formats impose a character length limit of between six and eight characters. You may find that the way the system truncates your tape name makes very little sense. You now have control to preserve the end of the tape name if that is where the most important digits of the tape library code exist. This is especially useful for tracking back tapes to a stock library.

PAL Lists

If you have a PAL project, be aware that the very first time you open EDL Manager, it may default to NTSC. You can change this to PAL under the Standards setting and it will stay that way. PAL lists become complicated when you are working in film and want to go back to the negative because of the two different methods of dealing with the difference between 24 fps (frames per second) and 25 fps. Sometimes you slow down the audio and sometimes you include extra frames in the video (that pesky 25th frame). The PAL Method 1 is for film projects in which video and audio are transferred together at 25 fps. PAL Method 2 is for when video and audio are captured separately and synced in Avid. If you have a video project without any matching back to film, then choose plain vanilla video PAL.

Matchback and 24-fps EDLs

Matchback is a method preferred by low-budget films or film projects that are primarily for video distribution, but will be projected sometime in the future. This is an option that will output a 30- or 25-fps project as a 24-fps cut list. Film transfer to video adds extra frames. There is no way for the video-based project to keep track of the A frame, so every outpoint in a matchback project can be accurate only to plus or minus a video frame in NTSC. The matchback is much more accurate in PAL. For many people, depending on the material, this is quite acceptable compared to the cost of conforming a negative first and then transferring to video or using a film composer. They get a video project at a high, two-field resolution that they can shop around or distribute, and when the funding begins to flow in, they can get the negative cut.

There is an easy way to make a 24-fps EDL for cutting negative or to import into a 24p finishing system like Symphony or DS. Highlight the Start column in the bin with all the sources used in the final sequence. Use Ctrl/Command+D to duplicate the column. You will get a pop-up dialog asking where you want to copy this information. Choose "TC 24" and the 30- or 25-fps start times will be converted to 24 fps. You can then use the 24-fps choice in EDL Manager to make your list. This is one of the least-known tricks for taking any film project to a 24-fps finish without using the Avid film options.

Using Symphony Universal at 24p allows the creation of both a high-quality output when working at 24 fps and a perfect negative cut list using FilmScribe™, a cut list utility. This will allow you to broadcast the finished project today and cut the negative when theater distribution is required. Using the Universal Editing option on Media Composer, you can output at 14:1 progressive (which looks pretty darn good for offline) and output the perfect list for cutting negative.

Conclusion

It should be clear by now that creating EDLs is rather complicated. It is also clear to the folks who receive these EDLs from

Avid systems that many people get it wrong. The amount of 9 AM thrash that is completely avoided by asking people to bring only a copy of their sequence bin more than pays for the EDL Manager software, so every postproduction facility should own a copy and have it quickly accessible. Of course, the translation to EDLs becomes less desirable every year as Avid solutions to finish using AAF are used by more facilities.

Here is a checklist of steps that should be taken before the linear online assembly:

☐ Simplify a copy of the sequence before you make the list. Do this based on feedback from the online editor.

☐ Format the floppy for the needs of the production house (either DOS or RT-11 or both).

☐ Use the EDL template sent by the postproduction facility or their faxed instructions for formats and options.

☐ Make any extra versions that might be important (audio only, video only, V1 separately, etc.).

☐ Use the same template to make a CMX 3600 version.

☐ Print the list twice. Make one for the editor and one for the producer. Include the source table in the printed versions only.

☐ Make a VHS copy of the sequence that you can control yourself in the suite.

☐ Don't forget to bring the source tapes!

INDEX

Printed and bound by CPI Group (UK) Ltd, Croydon, CR0 4YY

23/10/2024

01778255-0001